GREENLAND'S STOLEN INDIGENOUS CHILDREN

In this book, author Helene Thiesen recounts her experience of being removed from her family in Greenland as a young Inuk child, to be 're-educated' in Denmark and an orphanage in Greenland.

The practice of forcible assimilation of Indigenous children into colonial societies through 'education' has echoes in North America and Australasia, and the painful legacy of these practices remains under-acknowledged. In this poignant book, Helene recounts in detail the process of being taken from her family in 1951, aged seven, along with twenty-one other children, in the attempt to re-make them into 'model Danish citizens', in a social 'experiment' led by the Danish government and Save the Children Denmark. When the children returned to Greenland a year and a half later, they were sent to live in a Danish Red Cross orphanage, where they were forbidden to speak their native languages, and were compelled to adopt Danish language, culture and customs. With a detailed introductory analysis from Dr Stephen James Minton, who also provides the translation, Helene's account serves as a compelling and powerful testimony of a devastating colonial experiment.

Richly illustrated with forty photos to help to situate the reader, this book provides an invaluable case study for researchers and students in the fields of Indigenous Studies, Critical Pedagogy and Education, Psychology, European History, and Cultural Studies.

Helene Thiesen was one of the twenty-two Inuit children who, in 1951, were taken from their families in Greenland to be 're-educated' in Denmark. After a career in children's education herself, she has written a book about her experiences, which appears here in English for the first time.

Stephen James Minton is the translator and editor of this book. He is an Associate Professor in Applied Psychology at the University of Plymouth, UK, and a part-time Associate Professor in the Department of Education at the University of Southeastern Norway.

Routledge Studies in Indigenous Peoples and Policy

There are an estimated 370 million Indigenous Peoples in over 70 countries worldwide, often facing common issues stemming from colonialism and its ongoing effects. Routledge Studies in Indigenous Peoples and Policy brings together books which explore these concerns, including poverty; health inequalities; loss of land, language and culture; environmental degradation and climate change; intergenerational trauma; and the struggle to have their rights, cultures, and communities protected.

Indigenous Peoples across the world are asserting their right to fully participate in policy making that affects their people, their communities, and the natural world, and to have control over their own communities and lands. This book series explores policy issues, reports on policy research, and champions the best examples of methodological approaches. It will explore policy issues from the perspectives of Indigenous Peoples in order to develop evidence-based policy, and create policy-making processes that represent Indigenous Peoples and support positive social change.

Edited by Jerry White and Susan Wingert (The University of Western Ontario), this series considers proposals from across indigenous policy subjects.

Indigenous Intergenerational Resilience
Confronting Cultural and Ecological Crisis
Lewis Williams

Indigenous Health and Well-Being in the COVID-19 Pandemic
Edited by Nicholas D. Spence and Fatih Sekercioglu

Greenland's Stolen Indigenous Children
A Personal Testimony
Helene Thiesen, translated by Stephen James Minton

For more information about this series, please visit https://www.routledge.com/Routledge-Studies-in-Indigenous-Peoples-and-Policy/book-series/RSIPP

GREENLAND'S STOLEN INDIGENOUS CHILDREN

A Personal Testimony

Helene Thiesen

Translated, and with an Introductory chapter by,
Stephen James Minton

Routledge
Taylor & Francis Group

LONDON AND NEW YORK

Cover image: Helene Thiesen

First published 2023
by Routledge
4 Park Square, Milton Park, Abingdon, Oxon OX14 4RN

and by Routledge
605 Third Avenue, New York, NY 10158

Routledge is an imprint of the Taylor & Francis Group, an informa business

© 2023 Helene Thiesen and Stephen James Minton

British Library Cataloguing-in-Publication Data
A catalogue record for this book is available from the British Library

ISBN: 978-1-032-14936-3 (hbk)
ISBN: 978-1-032-14935-6 (pbk)
ISBN: 978-1-003-24184-3 (ebk)

DOI: 10.4324/9781003241843

Typeset in Bembo
by codeMantra

Frontispiece: Bishop Hans Fuglsang-Damgaard and his wife, who had been Barse-laj's foster parents, visited the Danish Red Cross Orphanage in Nuuk in the summer of 1953. From left, top row: Miss Blom, Bishop Fuglsang-Damgaard. Second from top row: Bodil, Big Kristine, Lene, Karen, Benze and Mrs. Fuglsang-Damgaard. Next rows: Eli, Eva, Agnethe, little Bodil and Ane Sofie, Albert, Marie, Regine, Little Kristine and me, Gabriel, Aron and Karl. Bottom row: Helge, Barselaj, Dorthe, Little Karl and Sâmo.

Front cover photograph: Me, on my eighth birthday – April 21, 1952. I had hidden myself at the back of orphanage, so that I could be away from the others, but one of the adults had run to find me, in order to take a photograph of the 'happy birthday child'. When they see it, people often say to me, 'How cute you are, in that picture!' But I want people to know how sad I was feeling at the time. I had been robbed of all of the most important things in my life. I had lost my father when I was six years old. I had been taken away from my mother and siblings when I was seven years old. And shortly before this photograph was taken, I had been told that I would not be permitted to even visit my mother and siblings, despite the fact that they lived only ten minutes away. My heart was bleeding – what did I have left? Only my doll, Tove, and my birthday present, Pluto – and no-one was going to take them away from me! When I got into bed, I hugged them tightly, feeling powerless and lonely, and crying inside myself: 'Why?' The photographer disturbed me in my feelings of grief and loss; at the time, I was wondering about whether it was worth celebrating my birthday at all, given that I was not allowed to live with, or even see, my mother and my siblings.

CONTENTS

FIGURES

NAMES OF PLACES IN GREENLANDIC AND DANISH

In this book, the old Danish names for the towns in Greenland have been used:

Danish		Greenlandic
Godthåb	=	Nuuk
Frederikshåb	=	Paamiut
Julianehåb	=	Qaqortoq
Sukkertoppen	=	Maniitsoq
Holsteinsborg	=	Sisimiut
Jakobshavn	=	Ilulissat
Egedesminde	=	Aasiaat

TRANSLATOR'S FOREWORD

Stephen James Minton

I began corresponding with Helene Thiesen in 2018, when I was inviting authors to contribute to a book that I was compiling and editing at the time, which was entitled, *Residential Schools and Indigenous Peoples: From Genocide via Education to the Possibilities for Truth, Restitution, Reconciliation and Reclamation* (Routledge, 2020a). A chance meeting that I had with some Greenlandic Inuit people in Copenhagen in 1997 meant that I had learnt *something* from first-hand sources about the lived experiences of Greenlandic Inuit people, with particular respect to the troubled relationship between Denmark and its former colony. (The Inuit people I met were enthusiastic about teaching this hitherto ignorant Englishman, and I was very willing to hear whatever they felt like telling me.) My learning in this area continued over the years, albeit accrued from reading various second-hand sources, which amounted to the very little that had been written in English about these matters. When I came to invite contributions to the *Residential Schools and Indigenous Peoples* book, I was keen that a chapter on Greenland should be included.

In the years before I made contact with Helene, I had seen her name in some of the aforementioned English-language sources, and on finding an article which was connected to previous BBC Radio broadcasts, I realised that I had heard something (again, second-hand) about Helene's life story already. This time, the source was my Mum (Rosemary Fox) who, as an avid BBC Radio Four listener, would often tell me about the more interesting programmes that she had listened to. My Mum, who died in May 2020 in the COVID-19 pandemic, was a true humanitarian, with a deep, personal sense of social justice. Accordingly, she was very touched by the programme – alternately saddened and outraged by what had happened to Helene, and deeply affected by her sense of Helene's personal warmth – as I am sure that any sensitive reader of Helene's work will be. In 2018, the author that I had previously engaged to write the chapter on Greenland for *Residential Schools and Indigenous Peoples* had to withdraw, and she suggested that I contact Helene in this respect, cautioning me that Helene was a retired lady, and might not have perfect English. None of that mattered; Helene and I found ways of working together on that chapter, as we have done on this book.

The chapter on Greenland in *Residential Schools and Indigenous Peoples* marked the first time that (a necessarily shortened version of) Helene's story had been published in an English-language academic text. What the reader now holds in their hands is much better – this is the full version of Helene's story, in her own words (in English translation). Helene and I have worked hard in bringing this book to fruition, and the personal and professional support we received from the series editor at Routledge, Professor Jerry White, and the editors Rosie Anderson and Helena Hurd, has been nothing short of incredible. Many people, it seems (including the reviewers of this book at its proposal stage, who I also wish to thank for their insights), now want to bring Helene's account of what happened to her as a child – the truth of which was kept hidden from everyone, including Helene herself, for forty-two years – to the Anglophone world. Amongst these have been my daughter, Anna; and my partner, Julie, who makes so many worthwhile things much more possible. For their encouragement of our efforts with this publication, and for so many other reasons, I send them both my love and thanks. I am extremely grateful to Lene Therkildsen, the owner and director of Milik Publishing (who published Helene's book in Danish and Greenlandic in 2011), for her agreement to this translation, and for her support of and interest in it. I would also like to express my sincere gratitude to the friends and colleagues who read and commented on my introductory chapter – Dr Hadi Strømmen Lile (University College of Østfold), Dr Julie Vane (British clinical psychologist), Dr Shawn Wilson (University of British Columbia Okanagan), and Arnbjørg Engenes, Dr Camilla Wiig, and Professor Willy Aagre (University of Southeastern Norway). Without the benefit of their suggestions for improvements, my chapter would have been considerably poorer.

On a technical note, the translation that follows was made from the Danish version of Helene's book, *For flid og god opførsel: vidnesbyrd fra et eksperiment* ['For Diligence and Good Behaviour: Testimony from an Experiment'] (2011, Milik Publishing). I have translated all of the Danish parts of the text into British English; words in Greenlandic are italicised, as they were in the Danish text, where Helene provided Danish translations of the Greenlandic. Over the course of this translation, I have provided some notes. Naturally, I have attempted to keep these intrusions to an absolute minimum, and have made them where Helene referred to things that would be familiar to Danes and Greenlanders, but in my view, less familiar to English-speakers. These include places in Greenland and Denmark, songs, stories, food and drink, and specific aspects of religious holiday celebrations. The photographs that appear in this text are all from Helene's private collection.

It is, of course, the primary authorship of this book that matters. Helene is one of only six survivors of the twenty-two Greenlandic Inuit people who shared her childhood experiences in a social 'experiment' of the early 1950s. Therefore, she is one of the few people who could give a first-hand account of these experiences, and as Helene is now into her late seventies, she is perhaps the only person who is ever likely to write this type of book. Amongst the many things I have learnt from my Indigenous researcher colleagues is the ethical value that if someone gives you their story, then you take on a responsibility for it. I have experienced that responsibility as the absolute privilege of working together with Helene on the translation and publication of her work in English. It has suited (and it still suits) many people to turn away from Helene's experiences, and from experiences like them. So as the translator, I thank the reader for having picked up this book; and I earnestly hope that my work on it, which I would like to dedicate to my late Mum, does the original author justice.

Exeter, England
March, 2022

FOREWORD TO THE DANISH- AND GREENLANDIC-LANGUAGE EDITIONS

Tine Bryld, 2011

I first met Helene Thiesen in 1990, when I was working on two books about the Greenlandic convicts sent to the closed prison, Herstedvester, for periods of indefinite duration. Helene said that she was a guardian for the mentally ill, convicted Greenlanders who, like the hereditary prisoners in the prison, had been sent to the Amts hospital in Vordingborg – indefinitely. I asked if there was a particular reason why she had taken on the difficult task of being a legal guardian. She replied, quite briefly, that as a child, she had been sent away to Denmark, and a year and a half later, had been sent back to an orphanage in Godthåb. I wondered. She had had her mother, and two siblings in town. Helene couldn't help me to find an explanation, except to say that she had felt very despondent, and rejected by her family. She had contact with a few other children from that time, and together we advertised for several of those who had been sent to Denmark, and back to the orphanage in Godthåb, to come forward.

After many attempts, and with the able assistance of the skilled staff of the National Archives, I was able to find the case of the twenty-two children who had been sent to Denmark in order to learn so much Danish that, after a year, they could travel back to the towns that they came from. The returned children were intended to be 'role models' for their peers who, after the new school reform, were to learn considerably more Danish. But that's not how things worked out. One factor was that a new school law had been passed in parliament on 27 May 1950, along with seven other laws for modernising Greenlandic society. Another thing was the way in which the legislation was implemented in the Greenlandic school system. A clear role was given to Save the Children, who were put in charge of finding the right foster families in Denmark for the children; and also to the Danish Red Cross, who had been given permission to build an orphanage in Godthåb. At that time, Greenland was plagued by great poverty, and many epidemics; child mortality was high in the years after the war.

Denmark was feeling pressure from the outside. Now the colonial power had to act, and to liquidate its colony, as had been the case with many other European countries. Hence, in 1953, Greenland was transitioned from being a Danish colony, to being a part of the Kingdom of Denmark. The twenty-two children were amongst the first to be 'Danified' – and in the worst sense of the word. They were not allowed to speak Greenlandic; they could not see their families; they were cut off from playing with Greenlandic playmates; and most of them could not understand or speak Greenlandic at all.

It was an experiment which was doomed to failure. It evolved into a series of on-the-hoof decisions, arrived at during meetings between the relief agencies and the Ministry for Greenland. Everyone was eager to help, and only a few protested when the matter was proposed to the Greenland National Council. At that time, the words of the Dane were almost law. A few years later, the process of sending Greenlandic children away to school in Denmark began. In total, this involved around 1,600 children; many of them can relate to feelings of need and longing, and not least of all, powerlessness. Some of these children's stories appear periodically throughout this book. For some, the experience was good; for others, it was a failure, and again, an experience of being split between two cultures.

In the 1960s, other experiments took place – anonymous adoptions that were a completely unknown concept in Greenland at the time; physically and mentally disabled Greenlandic people were sent to Danish institutions; and young Greenlandic people who had committed serious crimes were sent to Herstedvester and Vordingborg in Denmark, because there was no closed Greenlandic prison. These abuses can be explained by the fact that whilst Greenland was now officially part of Denmark, and Greenlanders should have had the same treatment as Danes, there were virtually no social services in Greenland itself. The Danish government had instead invested in making the infrastructure, port buildings, housing, and factories work as quickly as possible. This meant that education systems were slow to arrive, and that many vulnerable groups were not offered help in Greenlandic society. Thus, Helene Thiesen and her friends were the first group to put their bodies and lives into an experiment that no one knew about. The worst thing for me has been the fact that no one told the children and their families what it was all about. Only when I found the case in the National Archives were the children's stories returned to them.

I wrote the book, *I den bedsde mening* ['With the best of intentions'] in 1998, which in 2010 became the basis of the film *Experimentet* ['The Experiment'], which got new generations interested in what had really happened back then. In the journal *Social Kritik* ['Social Criticism'], no. 123/2010, the social policy of the 1960s is described; and now, Helene Thiesen's own account of how she and her friends were removed from their towns, villages, and families, sent to a foreign country, and split up amongst Danish foster carers is here. They became strangers, stripped of their identities. Back then, Helene and I talked about her writing about her family, the background of the grief that she has lived with for many years, and the failure that she felt. I knew that Helene could write so personally, and accurately, that one felt almost present in her grief and longing for her own family. The years have gone by, and Helene has been working constantly on her own text. I am convinced that it will mean something fundamentally valuable to all of those who shared their fates with her – those who were sent to Denmark in large numbers as children and who, as adults, have felt the rootlessness and emptiness.

Very few people have articulated their experiences at that time, and so it is of particular importance that the Greenlanders who remember those days write and talk about what they thought then, and what it meant to their lives and families. What was everyday life like? How did their families live? What was it like to have Danish teachers, and not to understand a word of what was being said? I think that Helene Thiesen will be joined by others in the years to come. Previously, for example, the artist Pia Arke, who unfortunately died far too young, wrote *Scoresbysundhistorier* ['Stories of Scoresbysund'], and more recently Ane Sofie Hardenberg and Pia Christensen Bang have published *Kampen for en far* [The Fight for a Father'].

Now we have Helene Thiesen's weighty tome, about the child who lost her family in the service of goodness. Without her husband, Ove Thiesen, and her children and grandchildren, this book could not have been written. They have always been there.

GREENLAND'S STOLEN INDIGENOUS CHILDREN AND THE SHADOW OF AN 'EXPERIMENT'

Stephen James Minton

Given the likely readership of this book, the purpose of this chapter is to help readers make connections between a context and a story with which they may be less familiar – respectively, the colonial relationship between Denmark and Greenland, and Helene's childhood experiences – and histories with which they are more likely to be familiar – e.g., the residential schooling of Indigenous children, in what are now (predominantly) Anglophone nation states.

Greenland and Its Colonisation

In August 2019, it was widely reported in the international news media that the then-US President, Donald Trump, was considering an attempt to buy Greenland. As the UK's *Guardian* newspaper's journalist, Martin Pengelly (2019) reported, Trump seemed to be seeking '…to tie the idea of a US purchase… to his own area of professional expertise, saying it would be "essentially a large real estate deal"'. Interestingly, with respect to answering the Google question that was most frequently suggested when the term 'Greenland' was entered into such a search at the time – 'Who owns Greenland?' – Trump's words were clear: 'Denmark essentially owns it'. He then went on to assert that this 'ownership' was '…hurting Denmark very badly because they're losing almost $700m a year carrying it. So they carry it at a great loss and strategically for the United States it would be very nice'. However, the response of the then-new Danish Prime Minister, Mette Frederiksen, to Trump's statements was equally clear: 'Greenland is not for sale… I strongly hope that this is not meant seriously' (Pengelly, 2019). Whilst it is tempting to locate some of the circumstances of this bizarre exchange in Trump's much speculated upon inner world, I believe that the question as to how it came about that any US President of 2019 would be attempting

DOI: 10.4324/9781003241843-1

to purchase a land inhabited primarily (85–90 per cent) by Greenlandic Inuit people, from the government of a European nation located almost 2,000 miles to the southeast, is very much worthwhile considering in its broader aspects.

The 2021 population of Greenland (in *Kalaalisutt* [Western] Greenlandic,[1] *Kalaallit Nunaat* ['land of the Kalaallit']) was estimated at 56,877, 85–90 per cent of whom were Greenlandic Inuit people (which included mixed-ethnicity people) (World Population Review, 2021). Greenland's non-Inuit population includes (in order of frequency) Danes and Danish Greenlanders (the descendants of Danish emigrants to Greenland), other Europeans (mostly Icelanders), North Americans, and other foreign nationals (chiefly, Thai, and Filipino) (Statistics Greenland, 2018). Whilst archaeological evidence exists of the region being populated by successive Paleo-Inuit cultures for the past five millennia, genetic studies indicate that today's Greenlandic Inuit people are the descendants of the Thule culture, who are generally believed to have arrived from Alaska in around 1300 CE (Grønnow, 2017; Sale & Potapov, 2010). What is absolutely certain is that for the last millennium, Greenland and its Indigenous population has experienced colonisation by European peoples, and despite limited home rule since 1979, and its official 'autonomous territory' status, remains part of the Kingdom of Denmark.[2]

The first European to found settlements in Greenland was the Norseman Eiríkr Þorvaldsson (Erik Thorvaldsson, or Erik the Red), who landed from Iceland in 982 CE, returning four years later to establish Norse settlements which are estimated to have numbered around 4,000 people at their peak, and seem to have persisted until around 1410 CE (Diamond, 2006; Sale & Potapov, 2010). However, for multiple reasons that remain contested by historians – see, for example, Diamond (2006) and Sigurðsson (2008) – these societies collapsed, and no evidence of Norse survivors was found in expeditions sent by the Danish king Christian IV between 1605 and 1607. Danish colonisation of Greenland began in 1721 when, having convinced the king of Denmark-Norway[3] of the benefits of acquiring Greenland, a Norwegian pastor named Hans Egede founded the settlement of Godthåb (now known as Nuuk, Greenland's capital). The fifteen years that Egede spent in Greenland were marked by his missionary activity (converting Inuit people to Lutheran Christianity), the establishment of other Danish settlements, and a smallpox epidemic (caused by the visit of a Danish girl) that wiped out one-quarter of the population (Sale & Popatov, 2010).

Petersen (1995, p. 119) characterised this colonisation as '…synonymous with mission and trade station', with the Lutheran missions and whaling stations in Greenland being managed by a single Dano-Norwegian enterprise known as the *Almindelige handelskompagni* ('General Trade Company'). In 1774, this was succeeded by a Danish enterprise known as the *Kongelige Grønlandske Handel* ('Royal Greenland Trading Department'), which managed the government of Greenland until 1908 and exerted an absolute monopoly on Greenlandic trade until 1950.[4] This being said, there was periodic trading contact between some Greenlandic Inuit populations and other Europeans who had commercial and territorial interests in the Arctic – notably the British and the Dutch (Petersen, 1995) – and whilst Norway had lost its offshore territorial claims in 1814, its commercial interests (fishing and whaling) in the region increased significantly, especially after independence. However, following the claiming by Norwegian fur-trappers of an unpopulated area in Greenland they called '*Erik Raudes land*' ['Erik the Red's Land'], in 1933, the International Court of the League of Nations ruled in favour of Denmark's sovereignty over all of Greenland (Sale & Potapov, 2010). Absolute certainty regarding this status was compromised by the occupation of Denmark by Germany during the Second World War, when Greenland was made a US protectorate, with a further treaty regarding the US's defence of Greenland being signed after the war. Overall, the United States established seventeen military sites in Greenland and began building the airport base at Thule in 1951.[5] It was in this year that Helene's tragic personal experiences of Danish neocolonialism, as evidenced in her recollections in the major part of this book, began.

'Northern Danes'

The years that followed the Second World War saw a number of European nation states – at that point, in economic ruin – either 'losing', or divesting themselves of, their former colonies. By 1981, the largest European empire, the British Empire – which at its 1920s peak had held sway over nearly one-quarter of the world's population, and occupied almost the same proportion of the Earth's land surface – had been reduced to fourteen 'dependent territories' and a Commonwealth of fifteen independent realms. In his 'wind of change' speeches of 1960, given in Ghana and South Africa, British Prime Minister Harold Macmillan recognised that the '…wind of change is blowing through this continent. Whether we like it or not, this growth of national consciousness is a political fact' (Watts, 2011).[6] Hence, the 'reduction' of the British Empire led to the political independence of many of its former colonies, some of whom retained an association as realms within the British Commonwealth. The Danish situation was rather different. In 1947, the United Nations had instructed Denmark to report on the development of Greenland and subsequently asserted that Greenland had to be decolonised (Heinrich, 2019). Under the new Danish constitution of 1953, Greenland's status as a colony was formally ended, and Greenland was made a constituent county within the Danish realm (Ehrlich, 2003; Fleischer, 1996; Petersen, 1995; Sale & Potapov, 2010)[7] – that is to say, it was entirely incorporated.

A good deal of the political discourse about Greenland in Denmark following these constitutional changes focussed on 'modernisation' measures, which would supposedly give Greenlanders – now to be considered as 'Northern Danes' – an 'equal' or their 'best' chance in the Danish Realm. Administratively, Greenland would be represented in the Danish *Folketing* [national parliament], and the 'modernisation' measures would include a campaign to reduce tuberculosis (the leading cause of death in Greenland at the time), making improvements to housing and the industrialisation of the Greenlandic economy. Heinrich (2019) asserted that the motivations behind the modernisation were threefold:

> Denmark wanted to continue to control Greenland and was thus willing to invest heavily in it; Greenlanders wanted to maintain the connection with Denmark as it gave them a feeling of security; and Denmark could boost its own status by being seen to help an under-developed society.
>
> *(p. 20)*

Significantly, these measures would be planned centrally in the Danish capital, Copenhagen, and their implementation would rely heavily on the 'import' of Danish 'expertise' and labour. Hence, as Petersen (1995, pp. 121–122) noted:

> ...no real change occurred, as Denmark for a long time administered the common human rights or civil rights in Greenland and continued to govern Greenland with the same civil servants and the same administrative body as before... modernisation made Greenland economically more dependent on Denmark than ever. The Danish staff in administration, and not least in education, introduced Danish ideas concerning economic activities and organisation. The means of attracting Danish staff to Greenland were economic, housing, and social privileges. This created a really visible discrimination between colleagues according to their Danish or Greenlandic origin.

Significant, too, was that the fact that the precise means by which the Greenlandic education system was to be reorganised were far from clear, although Danish language, culture and systems would be firmly in the ascendant. The use of Danish 'expertise' in the 'modernisation' of Greenland meant that the salaries of Greenlanders were much lower than those of Danes, who maintained that:

> ...it was not necessary to educate the Greenlanders because they were expected to be merely fishermen and plant workers. Greenlanders began to feel as if they were bystanders with no real influence... Modernisation in Greenland meant "Danification" which dictated that Greenlandic society be transformed into Danish society and meant that all things Danish were superior. The Greenlandic language was seen as unfit for the purpose of education, and schools were established in which the highest performing children were taught in Danish... Greenlanders did not imagine that all things Greenlandic would be eradicated. From a Danish perspective, the idea was to wipe the slate clean and start over with a new culture and a new language.
>
> *(Heinrich, 2019, p. 121)*

The Role of 'Education' in the Attempts to Forcibly Assimilate Indigenous Populations

It is true to say that, unlike what has been the experience of Indigenous populations in the Americas, no physical wars of occupation were mounted against the Greenlandic population by Denmark. Although – as we have seen – its claim to Greenland is rooted in the mires of Viking exploration, shifting Nordic political alliances, highly questionable missionary activity and commercial exploitation, and a contested League of Nations decision, Denmark generally proceeded as if it was legitimate and inevitable. Whilst not taking quite the same route as that of the early British colonisers in Australia – whose *terra nullius* approach obviated the very physical existence of the aboriginal populations – Denmark assumed control of, and the right to control, Greenland and treated the Indigenous populations as if (apart from constituting a source of cheap labour) they did not matter. So whereas in other parts of the world, it is possible to say that the '...genocide[8] perpetrated against Indigenous peoples via means of education accompanied, followed on from, and may be considered to have completed the grim task of, physical genocide' (Minton, 2020b, p. 3), given that no physical wars of occupation were made against the Indigenous Sámi in northern Scandinavia either, the historical positioning of Indigenous assimilation via education appears to have been somewhat different when one is considering the colonisation of Indigenous spaces by Nordic peoples.[9] Nevertheless, there are some strong systemic and psychological similarities across these processes, which we will now consider.

Previously, in a book I edited and contributed to – *'Residential School Systems and Indigenous Peoples'* (Minton, 2020a) – I sketched out four broad frameworks, or as I called them, 'theoretical touchstones', which reflected various concepts referred to by the other contributing authors, and informed some of the reflective material in the final chapter of the book (Minton, 2020c).[10] These were *'Indigenous as "Other"'* – the fact that Indigenous peoples have overwhelmingly been seen, understood, and deliberately socially/historically/politically positioned as 'Other' by Europeans and settler populations; *"Assimilation and Nation State Identity"* – as nation states have expanded, they have met their resultant 'security concerns' in the territory they claimed by positioning the land as a single state for a single people, into which any remaining Indigenous populations are to be assimilated; *'Educational Systems as Agents of (Cultural) Genocide'* – how the planning and implementation of the residential schools systems obscured and subverted truths at the time, and how such truths continue to be obscured and subverted and, even in the face of overwhelming evidence, outright denied; and *'The Residential School as a "Total Institution"'* – in the sense that Goffman (1961/1991) used the term; therefore, those institutions that 'permit' abuse and neglect to occur, and to remain concealed and obfuscated, and for culpable individuals, organisations, and nation states to position themselves as benevolent and philanthropic (Minton, 2020c). Of course, 'total institutions' other than residential schools for Indigenous children have existed and continue to exist,[11] and residential schools were just one of the many manifestations of the genocidal efforts and policies directed against Indigenous peoples by colonists. So in understanding the situation in Greenland, where (unlike, for example, Aotearoa, Australia, Canada, Norway, and the United States) no system of residential schools for Indigenous assimilation was ever established, the frameworks above will not entirely suffice. Nevertheless, for the 'Northern Danes' of Greenland, schools, school systems, and educational policies were important – both in what did happen and in what didn't happen.

Let us pick up one of the more challenging of the ideas that have recently been expressed; that is, that culpable individuals, organisations and nation states could, did (and still do) position themselves and their actions as benevolent and philanthropic. How is cultivating and maintaining this remarkable dissonance – genocidal intent and benevolence/philanthropy – even possible? Whilst the term 'total institutions' dates no earlier than the 1960s, the use of institutions of this type in the supposed 'reform' of those interred is, of course, much older – one could say that it has existed for as long as an element of 'reform' has been positioned as a constituent part of the incarceration of law-breakers. To take England as an example, the series of so-called 'Poor Laws', which date from the fourteenth century and the recovery after the Black Death, effectively criminalised destitution, vagrancy, and to some extents, nomadism, and culminated in the Poor Law Amendment Act of 1834, and the widespread establishment of workhouses (which had been developed over the previous two centuries) throughout the United Kingdom from the 1840s. 'Reformatory' schools' were established at around the same time, according to the model of the founder of the first of their kind, Mary Carpenter (Lynch & Minton, 2016).[12] Conditions in workhouses were deliberately austere; poor diet, poor or non-existent education (for children), harsh discipline, mismanagement, and the separation of families were general features (see Higginbottom, 2012, for review).[13] The people in workhouses and the reformatory schools were thereby blamed for their criminalised poverty, but the 'reformers'' stated hopes were that such 'moral failures' could be ameliorated. In much the same way, attempts would be made to 're-form' the Indigeneity of children in residential schools the world over (see Minton, 2020a, for review). Hence, it is not the case that residential schooling was invented from the 'ground up' by settler populations – rather, those populations adopted and applied ideas, systems and institutional types that were already long-standing in Europe. So, the 'means' of 'reformation' were well-established – but where did Europeans and settlers assume the 'right' to implement these means from?

In medieval Europe, the idea of the 'Great Chain of Being' – a continuous progressive chain of non-human living forms, from beast to man, and thence through to the angels and God himself – was a familiar representation of the Christian view of Creation. So when European 'explorers' began to encounter non-European peoples, they inevitably positioned them lower on the 'Great Chain' than people of their own kind – that is to say, as lower ranks of humanity, and closer to the animals than they themselves were to God. Indeed, prior to the issuing of the *Sublimus Deus* Papal Bull (1537), which declared the Indigenous peoples of the Americas to be rational beings, with souls, and therefore condemned and forbade their enslavement, the *Inter Caetera* Papal Bull (1493) was generally (and conveniently) interpreted by the Castilian monarchs, and the conquistadors, as having divinely ordained their rights of conquest in these lands, and over its inhabitants who were deemed to lack souls, and could therefore be enslaved and murdered with impunity (see Charles & Rah, 2019; Horst, 2020; Stannard, 1992).

Whilst the Reformation and the seventeenth-century European religious wars caused great changes in certain areas of European thought, this swing towards apparent 'rationalism' and 'science' had the effect of further facilitating the continued and seemingly intractable 'Othering' of Indigenous peoples. The widespread, yet entirely erroneous, view of evolution as producing progression, rather than (as it does) diversity meant that the supernatural top rungs of the 'Great Chain' were effectively lopped off in the common European mindset, but the pre-existent gradation of human and non-human living forms remained intact. Coupled with Social Darwinism, this viewpoint provided Europeans with the opportunity to falsely 'biologise' cultural, historical, and economic differences (see Gould, 1997), and to develop (pseudo-) 'scientific', rather than theological, bases to their assumptions of inherent superiority over non-European peoples. It appears that regardless of whether the Western (and Westernised) mindset is informed by the Semitic religions, Social Darwinism or, as it is today, by Eurocentric patriarchal neoliberal capitalism, the positioning of peoples of colour, or those with any characteristic that can be used to identify them as 'Other', and thereby maintain their marginalised status, remains the same.

Furthermore, under the Social Darwinist view that 'savage' societies inevitably give way to 'civilised' societies, a 'philanthropic' position could be asserted (by factoring in the understanding that 'savage' culture is not necessarily 'in the blood', but rather 'in the upbringing') holding that hitherto 'uncivilised' people could, and even should, be given a chance to adapt to 'civilised' culture and take their places in the new society. This 'generosity' on the part of 'civilised' European and settler cultures was bolstered by an idea of 'best interests' – that there were some groups of people, including Indigenous peoples, who thrive best under strict and responsible supervision. I would argue that such notions of superiority and entitlement underpin the apparent 'certainty' of European and settler populations in their notions of the desirability, likely efficacy, sense of moral 'correctness' – and even 'generosity' – in the use of schooling in the forcible assimilation of Indigenous populations. Furthermore, these beliefs became so entrenched and pervasive as to render their holders unable to recognise, much less to acknowledge, that their presence and consequences (including demonstrable institutional neglect and abuse) had any negative impact on the lives of Indigenous populations.

In Helene's story of her childhood, a key figure is Miss Dorothea Bengtzen, who was known as 'Benze' by the Greenlandic children (including Helene) who resided at the orphanage that she managed. As I read it – and readers will, of course, be able to decide for themselves if they share my view – one of the most consistent aspects of Benze's character is her unfailing belief in the superiority of Danish (over Greenlandic) culture. This is evidenced in a variety of areas, not least in her choice of friends and associates – all of whom seem to be relatively affluent Danes, of some type of local importance – which becomes obvious to the Greenlandic orphanage children. This same assumption of Danish cultural superiority leads to many examples of unthinking emotional cruelty towards the children – or critically, interactions that Helene's words reveal to us were experienced, as the child that she then was, as psychological abuse. I believe that in all probability, Benze would have either been unaware of the impact of her assumptions and actions or, had she been ever been challenged on these, would undoubtedly have disavowed them. A concrete example of these types of assumptions and actions is seen in the run-up to the first Christmas after the children have moved into the orphanage. Helene is expecting to go home to her mother's house for Christmas and tells Benze that she is looking forward to showing her mother the cardboard Christmas tree that she (Helene) has made. Benze bluntly informs Helene that she is mistaken – the orphanage is her home now, and she can instead look forward to a real Danish Christmas, with all the trimmings. Importantly, Benze seems capable of kindness; for example, when she finds some of the orphanage girls (again, including Helene) out of bed after lights out, and dancing around one of the cardboard Christmas trees, she pats them on the back and bids them a good night.

The secondary positioning of Greenlandic language, culture, family, and identity was a constant feature of orphanage life, and the Danification of the children was vigorously pursued. Regardless of their linguistic competencies before they entered the orphanage, they emerged from it as solely Danish-speaking. According to the schools that they went to, some of the children learnt foreign languages, such as German and English, but none of them were allowed to speak, retain, or learn Greenlandic. This secondary linguistic and cultural positioning is evident in Helene's account of the shadowy and marginal existence of the *kiffaks* (who were poorly paid Greenlandic maids) amongst the orphanage staff. There are many moments of solidarity between the *kiffaks* and the Greenlandic children in the orphanage, and Helene provides an account of some of the orphanage girls trying to (re-)learn some Greenlandic words from one of the *kiffaks,* a woman named Birthe. Benze had quickly arrived on the scene, and Helene recalled her chastisement of Birthe: 'I never want to hear you trying to teach these children Greenlandic again. They are Danish-speaking, and they must continue to be so. Is that understood?'. Helene also recalled her feelings of emptiness and potential loss (the fear of not being able to speak to her Greenlandic-speaking mother and siblings in future), her anger at Benze, and her sympathy with Birthe (later that day, at dinner, Helene refused to meet Benze's eye-contact, but deliberately tried to make eye-contact with Birthe). Eventually, however, some of the orphanage children learned to take advantage of, and sometimes make fun of, the staff members who could not speak Danish well. As was the case in residential schools for Indigenous children, every effort was made to teach the orphanage children to disdain what they themselves were, and to instead adopt the cultural models of the colonisers as the normal, desirable, and aspirational position. Any gestures of refusal on the part of the children to fully internalise such 'teaching' – in other words, failing to learn these 'lessons' – must be considered as acts of resistance, under conditions of complete disempowerment. Undertaken as they were by interred children, some of these gestures of refusal may look small, and more often than not, they were ineffective; nonetheless, these actions were as heroic as any of those that have been made by those resisting personalised experiences of extinction.

Some Notes on Denmark's Colonial Legacies in Greenland

As the reader will find, Helene finishes the bulk of her autobiographical material in the summer of 1960, when she was sixteen years old. I feel that it might be helpful to the reader to know about some things that have happened, in terms of Greenland's relationship with Denmark, over more recent decades. After a referendum held on January 17, 1979, in which 70.1 per cent of people (of the 63.3 per cent of registered voters who turned out) voted for greater autonomy from Denmark, the Greenlandic Home Rule Act was passed on May 1, 1979. This gave Greenlanders the right to elect their own Parliament, with sovereignty and administrative responsibility in the areas of education, environment, fisheries, and health (but not in the areas of civil rights, defence, finance, justice, or national security). In a subsequent referendum, held on November 25, 2008, 75.5 per cent (of the 72.0 per cent of registered voters who turned out) voted for Greenlandic self-government, and the Self-Government Act took effect in June 2009. The most significant aspects of this were Greenland taking control of its mineral and oil rights (which had previously been co-managed with Denmark), and the possibility for full independence, provided that this measure was supported by a referendum (Kuokkanen, 2017). A poll conducted in 2016 showed that 64 per cent of respondents favoured this (Skydsbjerg & Turnowsky, 2016); however, a poll conducted in 2017 showed that /8 per cent were opposed to independence if it resulted in a lowering of living standards (Bjerregaard, 2017). Denmark continues to pay an annual subsidy to Greenland, which has usually amounted to about 30 per cent of Greenland's GDP (Kuokkanen, 2017), and should independence be declared, this subsidy will cease.

As well as this financial leash, the effects and legacies of Danish colonisation have been, and still are, evident in public health in Greenland. Historically speaking, and as was the case in the Americas, Indigenous people in Greenland had no resistance to the infectious diseases (e.g., influenza, measles, smallpox, tuberculosis, and whooping cough) which Europeans brought with them, and these had similarly devastating effects. Inuit people in Greenland remain particularly at risk of infectious diseases such as hepatitis, meningitis, pneumonia, and sexually transmitted diseases, and chronic diseases such as specific forms of cancer, diabetes, heart disease, hypertension, obesity, and stroke. With the traditional Greenlandic diet, and the hunting and outdoor lifestyle being replaced by Western foods and sedentary habits, smoking and the use of alcohol have become prevalent, the latter of which is implicated in many incidents of abuse, accidents, and interpersonal violence (Bjerregaard, Kue Young, Dewailly & Ebbesson, 2004), and gambling is also a problem (Larsen, Curtis & Bjerregaard, 2013). Perhaps above all, at an incidence rate of 83 per 100,000 of the population between 1985 and 2011 (World Health Organisation, 2011), the suicide rate in Greenland has been by far the world's highest.[14] It has been calculated that one in five people in Greenland has attempted suicide (Bannister, 2010). Most of those who kill themselves in Greenland are boys between the ages of 15 and 19 years old (Bannister, 2010), who tend to do so by violent means (e.g. hanging or shooting) with seasonality, alcohol abuse, and impulsivity (Björkstén, Kripke & Bjerregaard, 2009) and psychological (depression) and social (familial and domestic conflict, poverty) (Hersher, 2016) factors having been implicated or demonstrated as risks.[15]

The Persistence of Denial and Obfuscation...

Hayner (2010) noted that in the fifteen years that followed the internationally well-known truth and reconciliation processes in South Africa, over twenty truth and reconciliation commissions were established around the world. Whilst perhaps not quite amounting to the adoption of such processes as a global 'panacea' in the attempt to deal with conflict and post-conflict situations, this must at least be reckoned as a broad-scale adoption, heralded by an apparent 'success story' in post-apartheid South Africa. Whilst many have cast doubts regarding the appropriateness and likely efficacy of such processes in addressing the colonial and settler-colonial experiences of Indigenous peoples,[16] measures of these types have been attempted in Greenland, in the same way that similar processes have been attempted in Australia (see Norman-Hill, 2019 for review) and Canada (Truth and Reconciliation Commission of Canada, 2015) and (at the time of writing) are currently being attempted in Norway, and have been being called for in the United States for some time, although – significantly – what are being urged now are processes of truth and *healing*, rather than truth and *reconciliation* (National Native American Boarding School Healing Coalition, 2021).

Greenland established a Reconciliation Commission in 2014, but hopes for the success of meaningfully investigating the effects and legacy of colonialism in Greenland (Rud, 2017) and initiating a process of reconciliation with Denmark (Andersen, 2014) – which were amongst the aims that the then-Greenlandic Prime Minister, Aleqa Hammond, had for the process – can only have been mixed from the outset, given comments made by Hammond, and her then-Danish counterpart, Helle Thorning-Schmidt at the time. Whilst Hammond stated that 'Reconciliation is very important on a path where Greenland strives for independence... For a country that moves toward greater autonomy and independence, its people need to know about their own story' (in Jacobsen, 2015), Thorning-Schmidt characterised efforts towards reconciliation as being a purely Greenlandic aspiration, rather than a mutual one: 'We do not need reconciliation, but I fully respect that it is a discussion that occupies the Greenlandic people. We will follow the discussion carefully from here' (in Therkildsen, Olsen, Mathiassen, Petrussen & Williamson, 2017, p. 16; translation mine). Perhaps predictably, the final report of the Reconciliation Commission in Greenland included the recognition that, '...from being a reconciliation commission that would also bring about reconciliation between Denmark and Greenland, the Commission's task was narrowed to being a reconciliation process in Greenland alone',[17] and that its work should rather, therefore, '...be considered a commission of truth'[18] (Therkildsen, Olsen, Mathiassen, Petrussen & Williamson, 2017, p. 17; translations mine).

So much for Denmark's participation to date in larger processes of truth and reconciliation with its former colony, and its population of so-called Northern Danes, then. But what, specifically, of Helene, and the other survivors of the 'experiment' of the 1950s? At the time of writing, there are just six of the twenty-two still living, and when I asked Helene about what had happened to the survivors, she told me that

> ...some had died young; many had experienced difficulties in partner relationships; and some had experienced homelessness. They could not live as other people did, as they were never allowed to return home – to where they were born; to their parents, siblings and other family members – and had therefore become strangers in their own country.
>
> *(Minton & Thiesen, 2020)*

As she had explained elsewhere, 'Some just broke down. They lost their identity and they lost their ability to speak their mother tongue and with that, they lost their sense of purpose in life' (Otzen, 2015). It was in 1996, when the news of *why* Helene had been taken away from her family was finally broken to her, that the term 'experiment' first became attached to this set of experiences. Significantly, this news came from a Danish writer (Tine Bryld), and not a government, nor any form of state or charitable source. Helene recalled, 'She (Bryld) called me up and said, "Are you sitting down? You've been part of an experiment"' (in Otzen, 2015). This term (the 'Experiment') has generally been used since (not least by Helene herself, in the subtitle of her own autobiographical work, and also as the title of a feature film (Friedberg, 2010) and seems to fit in some important ways, particularly if 'experimentation' is seen as a reckless playing with vulnerable lives by powerful authorities. But can the process of transferring the twenty-two Inuit children back and forth between foster care in Denmark and an orphanage in Greenland be seen as an 'experiment' in other senses of the word?

In the first place, and despite the debates at the time about the language of instruction to be used at various levels of schooling in Greenland, this 'experiment' cannot be positioned as a piece of formal educational research into the composition and modes of delivery of school curricula. It takes no great effort on the part of any reasonably intelligent person to think of the myriad of ways in which such data could have been obtained, should it have been required, none of which would have involved sending Greenlandic children away to Denmark or incarcerating them in an orphanage on their return. There is then the question of documentation, both at the time and since. The fact of the matter is that Tine Bryld had an exceptionally difficult task in compiling the scant official records that still exist regarding Helene and the other children's experiences; she stated that it was only '…after many attempts, and with the able assistance of the skilled staff of the National Archives, [that] I [she] was able to find the case of the twenty-two children' (Bryld, 2011; translation mine). As the reader will find out from reading Helene's testimony, and whilst being a feature that many people find shocking and surprising, the Danish Red Cross and Save the Children had clear, unequivocal, and key involvements in these events. Helene received a letter from the Danish Red Cross in 1998, stating that the organisation now 'regretted its role', and Save the Children apologised in 2010 and 2015.[19] The internal investigation by Save the Children showed that some of the documents had 'disappeared' and that they could not rule out the possibility that they had been deliberately destroyed. Significantly, in the historical study of the 'experiment' commissioned by the Danish and Greenlandic prime ministers, which involved a review of all of the available and existent official documentation, Jensen, Nexø & Thorleifsen (2020)[20] observed, 'We have not found any evidence that the "experiment" was at any time the subject of an actual evaluation by the organisations and authorities involved, or on the part of politicians'. Additionally, much as the so-called experiments done in school laboratories are, in fact, *demonstrations* – there is no doubt that, if sufficient care is taken, the rehydration of the anhydrous copper sulphate powder will produce the familiar blue copper sulphate crystals – the results of the 'experiment' in which Helene and the other children were unwilling and involuntary participants should have, or at the very least could have, been known in advance. It is impossible, therefore, to see these 'results' as anything other than being one and the same as the 'aims'.

Despite the unclear formulation of research questions, the lack of control measures, the poor or non-existent documentation, and the conflation of aims and results, there are, however, some good reasons to continue the use of the term 'experiment' with respect to this set of experiences. In the first place, the term has been used by those themselves who were involuntarily involved, and its use has been established more generally, too – in Danish, it is not uncommon to see those twenty-two people referred to as the '*(Grønlandske) eksperimentbørn*' ('(Greenlandic) experiment children') (see, e.g., Ahrens & Møller, 2021; Ritzaus Bureau, 2022). Second, it is to be recognised that the identification of a set of intelligent Greenlandic children from under-resourced families, in order to deliberately manipulate their lives, identities, and culture towards the model of the colonisers, *did* sound like a laudable and worthwhile set of aims to those involved. Regardless of the face-saving and responsibility-limiting possibilities that such a revisionism might permit, this was not an 'educational' experiment that somehow 'went wrong'. Greenlandic children had been sent away to school in Denmark before, and 1,600 Greenlandic children would be similarly sent away in the years following the 'experiment' (Bryld, 2011). What was different in Helene and her fate siblings' case – and to borrow the language of experimentation – was that an additional, and quite possibly, a main 'variable' was involved, i.e., the deliberate separation of Indigenous children from their families and communities, in order that efforts at assimilation via education could be further facilitated.[21] So if we are to continue to use the term 'experiment', let us be clear about what was under consideration – which was the application of what were thought to be the most effective methods of neocolonial Indigenous assimilation. Specifically, in this case, mired as it was in '…on-the-hoof decisions, arrived at during meetings between the relief agencies and the Ministry for Greenland' (Bryld, 2011), it was aimed at producing the seeds of future generations of little 'Northern Danes'. Hence, it was a 'social' experiment, which led to disastrous social consequences. Here, too – on the subject of the 'results' as social consequences – Jensen, Nexø & Thorleifsen (2020)'s summaries were starkly clear:

> With the knowledge that is available today, however, it is quite clear that the "experiment" in the long run did not lead to the desired results… The children did not function as bridge builders between Denmark and Greenland. Their education and employment did not match the high expectations they had, and which the attempt to select particularly intelligent children, and the efforts towards stimulating their use of the Danish language and supporting their schooling, were aimed at. Most left their country, and thus did not get the place in its development and future that had been imagined…The adulthoods of a significant number of the twenty-two children were characterised by extensive social and/or personal problems…lives marked by alcohol or other substance abuse, and in some cases, by mental illness, hospitalisations, suicide attempts, petty crime, periodic homelessness and vagrancy. This coincides with the fact that more than half of the children did not live until they were 70 years old. The majority – even of those who do not seem to have struggled with such problems – came to live life without close ties to their biological families, and some of them did not start their own families, either…of the half of the children who did manage to build secure lives for themselves, and completed a vocational or academic education or obtained employment, many of them were left to fend for themselves – they were no longer the institutions' or the authorities' problem.
>
> *(pp. 71–72, translation mine)*

...and the Efforts towards Something Better

As Bryld (2011) and Jensen, Nexø & Thorleifsen (2020) found, and as the reader will find that Helene experienced, the 'experiment' was shrouded in considerable secrecy at the time, and remained so for generations. Politically speaking, calls from the Greenlandic government for an independent investigation into the 'experiment' were backed by the Danish Social Democrats when that party was in opposition, but dropped when they were in government (2011–2015).[22] Lars Løkke Rasmussen, a Liberal who had served terms as prime minister before (2009–2011) and after (2015–2019) Thorning-Schmidt, said in 2009 that the events had to be understood according to the thinking of the time. Rasmussen said that the intentions amongst those involved had been good, and:

> History cannot be changed. The government regards the colonial period as a closed part of our shared history. We must be pleased with the fact that times have changed.
>
> *(in Rud, 2017)*

Rasmussen therefore refused to apologise on behalf of the state (even when the example of the official Canadian apology to Indigenous peoples was put to him in 2010) or to entertain the possibility of compensation (Jensen, Nexø & Thorleifsen, 2020). Karla Jessen Williamson, a Greenlander who was part of the Greenlandic Truth and Reconciliation Commission, sent to high school in Denmark, and is now an Assistant Professor in Educational Foundations at the University of Saskatchewan, was very clear in her characterisation of the ignorance and indifference that persists in Denmark around the 'experiment', and indeed, about Greenland in general[23] (CBC, 2022):

> There's only ten per cent of the Danish population that has been sort of receiving some kind of knowledge about the twenty-two children that were forced into becoming little Danes. But I would say that, generally speaking, that only five per cent of the Danish population would know anything about the history of Greenland. The Danish school system has been absolutely denying the history of Greenland as being part of their own history. I see that as being very problematic. And the Danes have been very, very good at looking at themselves as being very good colonisers, compared to many other nations that we know in the world. And this kind of nationalistic look at themselves prevents them from seeing themselves as being a colonising nation.

However, it seems that through the efforts and actions of some journalists and broadcasters, both in Denmark (see, for example, Ahrens & Møller, 2021; Attardo, 2020; Matthiessen, 2021; Mørck, 2020) and internationally (see, for example, BBC News, 2020, 2021; CBC, 2022; Otzen, 2015; Poulsen, 2021;

Rasmusson, 2019; Weichert, 2021), as well as those of some activists, authors, film-makers, lawyers, and certain politicians (see Jensen, Nexø & Thorleifsen, 2020, for review), this ignorance may be beginning to shift. It was the publication of Tine Bryld's book '*I den bedste mening*' ('With the best of intentions', 1998) that first broke this story in Denmark. The feature film, '*Eksperimentet*' ('The Experiment'; Friedberg, 2010), which depicted the lives of the children at the orphanage in Nuuk appeared later, as did two survivor testimonies – Helene's '*For flid og god opførsel: Vidnesbyrd fra et eksperiment*' ('For Diligence and Good Behaviour: Testimony from an Experiment', 2011), and Carla Lucia Knakker-gaard's '*Et godt liv – trods alt! Tvangsflyttet fra Grønland til Danmark*' ('A good life – after all! Forced relocation from Greenland to Denmark', 2012).[24] And as we have seen, the Greenlandic Reconciliation Commission operated (without Danish involvement) between 2014 and 2017 (Therkildsen, Olsen, Mathiassen, Petrussen & Williamson, 2017).

A Greenlandic member of the Danish Parliament, Aaja Chemnitz Larsen, tabled a resolution on February 8, 2019 instructing the government to apologise '...to the persons who were removed from their families in Greenland in 1951 and sent to Denmark as part of a social experiment' (Jensen, Nexø & Thorleifsen, 2020). At around the same time, at a meeting between Danish Prime Minister Lars Løkke Rasmussen and the Greenlandic Prime Minister, Kim Kielsen, there was an agreement to commission a historical study into the case, which was delivered in November 2020 (Jensen, Nexø & Thorleifsen, 2020).[25] The following month, Prime Minister Mette Frederiksen – who had succeeded Helle Thorning-Schmidt as leader of the Social Democrats in 2014, and become Denmark's second female, and youngest ever, prime minister in June 2019 – reported that she had sent a letter to each of the six children still alive with '...an unreserved and long-awaited apology on behalf of Denmark', and stated that:

> I have been following the case for many years and I am still deeply touched by the human tragedies it contains...We cannot change what happened. But we can take responsibility and apologise to those we should have cared for, but failed to.
>
> *(Attardo, 2020; BBC News, 2020)*

Spring, 2022

The above written apology notwithstanding, the case for compensation (see Ahrens & Møller, 2021; BBC News, 2021; Matthiessen, 2021; Poulsen, 2021) did not run smoothly. The Danish state, through the Ministry of Social Affairs and the Elderly, rejected a claim made by the lawyer Mads Pramming for compensation of DKK 250,000 (ca. £28,000/€33,500/CA$47,000/US$37,000)[26] for each of the six survivors and redirected him to the courts. Pramming was in no doubt that the state had a direct responsibility in the case and that it had violated the human rights of the children, and therefore was fully prepared to direct his claim towards the Prime Minister's Office, as the bodies that started the 'experiment' have since been closed down, and their responsibilities have been transferred there (Matthiessen, 2021). However, the case was settled in the amounts above in February 2022 (France-Presse, 2022).

Frederiksen invited the six surviving members of the twenty-two Inuit children who had been taken from Greenland to Denmark to a special event in the Danish capital, Copenhagen, on Wednesday, March 9, 2022. In a face-to-face personal apology, Frederiksen told the survivors (France-Presse, 2022; Vaaben & Borberg, 2022):

> What you were subjected to was terrible. It was inhumane. It was unfair. And it was heartless. We can take responsibility and do the only thing that is fair, in my eyes – to say sorry to you for what happened.

The event seemed to me[27] to be far more meaningful than a set of politicians' speeches (the Prime Minister of Greenland, Múte Bourup Egede, also gave a speech, as did Eva Illum, one of the six survivors). It was a two-hour event, at which many tears were shed, songs sung, and Frederiksen's personal invitation and involvement in this face-to-face apology, and her spending time with the six survivors and their families, was deeply appreciated. As Eva Illum put it in her speech, 'Nothing had happened until now and it's you, Mette, who took the initiative to set up a commission two years ago' (Figure 1).

In another highly appreciated action, Frederiksen repeated the event and the face-to-face apology at a special meeting in the Greenlandic capital, Nuuk, on Tuesday March 16, 2022 – the place where Helene, who gave a speech at the meeting, had been taken from some seventy-one years previously (Figure 2).

In pondering the potential longer-term significance of the award of compensation in February 2022, and the face-to-face apologies of March 2022, I have found myself returning to some eloquent words spoken on a different continent, regarding a different struggle – those of Malik el-Shabazz (Malcolm X) who, in a television interview of 1964, said:

> If you stick a knife in my back nine inches and pull it out six inches, there's no progress. If you pull it all the way out, that's not progress. Progress is healing the wound that the blow made. And they haven't even begun to pull the knife out, much less tried to heal the wound. They won't even admit that the knife is there.

In the case of the 'experiment', Denmark (or at least, and at last, Denmark's current Prime Minister) seems to have begun to acknowledge that the knife is there and to pull it out. But with these events being so recent, and the 'great Danish ignorance' (see endnote 23) being so widespread, at the time of writing it

FIGURE 1 Helene Thiesen (front), with (left to right) Asta Søholm, Laila Hansen, Stephen James Minton, Anja Otten, and Mette Frederiksen (March 9, 2022)

FIGURE 2 Helene Thiesen's speech in Nuuk (March 16, 2022)

can only remain to be seen if, and to what extent, progress towards healing the wound will be made.

It is to be acknowledged that very few people set out to knowingly do the 'wrong thing'.[28] However, experiences in Greenland such as Helene's, and indeed across the entire Indigenous world, have evidenced time and again that great damage has been done by people who *thought* that they were doing the 'right thing'. Furthermore, major impediments to addressing the wrongs of the past – and their legacies in the present – are raised in the fantasies of Europeans, and settler populations, that their predecessors *did* the 'right thing' or, that whilst their actions might have been 'questionable' or even outright wrong, we would all do best to 'put the past in the past', perhaps somehow 'reconcile', and essentially forget about them, and that this 'forgetting' should be reflected in how we think, talk, and educate future generations about the realities of the past and present. This neoliberal consensus on 'agreed' amnesia is reflected in a myriad of places – yes, in Lars Løkke Rasmussen's afore-mentioned comments, but also in settlers casually standing over, or loudly insisting on their 'right' to don 'Native' attire in Hallowe'en costumes; their appropriation, most often in grotesquely distorted forms, of Indigenous spiritual practices; the continued use of Native mascots in sports teams, especially in North America; the disavowal of, and objection to, the correct use of the term genocide when describing colonialism, and the missing and murdered Indigenous women and girls; their muted, non-existent, or outright hostile responses to the raising of the inhuman scandals implicit in the attempt to assimilate Indigenous peoples via residential schooling, forcible adoptions, and to decolonisation, and demands for land return; and their sheer non-engagement with the legacy effects and contemporary manifestations of colonial processes, substituting instead attitudes of victim-blaming. Effectively, what these historically recent interlopers are saying, to the populations who have acted as guardians of these lands for millennia (or tens of millennia) – lands that settlers now occupy, via genocidal processes of colonisation – is, 'Well, that was all a long time ago, and whilst there were probably a few bad apples back then, those bad things were never done on purpose. I'm sure that my people were at least trying to do the right thing by yours. So what are you, who wasn't even born at the time, still complaining for now?'

The obvious answer to such a question – framed as it is in its repulsive, woefully, wilfully and deliberately cultivated ignorant pseudo-understandings – is that it is *always* wrong to attempt to destroy peoples and individual people. I would argue that in our own times when our all-too fragile notions of democracy are daily giving way to patriarchal populism, when the message that we mere citizens consistently get from the empowered is couched in the neoliberal view that 'there is no alternative', we find ourselves in the repugnant position of having to assert such obvious answers with increasing frequency and urgency. One way in which we can remind ourselves of the necessity of, and to bolster our efforts in, doing so, is to truly hear, or to read about, the lived realities of people like Helene. Therefore – and also, as I am conscious of having taken up a not inconsiderable amount of space already – it is my great pleasure to turn the reader's attention towards Helene's own words and story.

References

Adams, D.W. (1995). *Education for Extinction: American Indians and the Boarding School Experience 1875–1928.* Lawrence, KS: University Press of Kansas.

Ahrens, K. & Møller, E.-M. (2021). Grønlandske 'eksperimentbørn' kræver erstatning fra den danske stat. [Greenlandic 'experiment children' demand compensation from the Danish state]. *DR Nyheder,* November 22nd.

Andersen, A.N. (2014). *The Greenlandic Reconciliation Process: Project Description.* Available on-line: http://www.martinbreum.dk/wp-content/uploads/2014/12/Carlsbergansøgning-2014-EV.pdf [Accessed March, 2022].

Attardo, D. (2020). Mette Frederiksen siger undskyld til 22 grønlandske børn. [Metter Frederiksen says sorry to 22 Greenlandic children]. *Ekstra Bladet,* December 20th.

Bannister, M. (2010). Singing to end teen suicide in Greenland. *BBC World Service,* December 7th.

BBC News (2020). Denmark apologises to children taken from Greenland in a 1950s social experiment. *BBC News,* December 8th.

BBC News (2021). Greenland's Inuit seek Denmark compensation over failed social experiment. *BBC News,* November 23rd.

Bjerregaard, M. (2017). Redaktør: Grønlændere vil ikke ofre levestandard for selvstændighed. [Editor: Greenlanders will not sacrifice standard of living for independence]. *Danmarks Radio,* July 27th.

Bjerregaard, P.; Kue Young, T.; Dewailly, E. & Ebbesson, S.O.E. (2004). Indigenous health in the Arctic: An overview of the circumpolar Inuit population. *Scandinavian Journal of Public Health,* 32: 390–395.

Björkstén, K.S.; Kripke, D.F. & Bjerregaard, P. (2009). Accentuation of suicides but not homicides with rising latitudes of Greenland in the sunny months. *BMC Psychiatry,* 9: 20.

Brave Heart, M.Y.H. & DeBruyn, L.M. (1998). The American Indian holocaust: Healing historical unresolved grief. *American Indian and Alaska Native Mental Health Research: Journal of the National Center,* 8(2): 56–78.

Bryld, T. (1998) *I den bedste mening.* [With the best of intentions]. Nuuk: Atuakkiorfik.

Bryld, T. (2011). Forord. [Foreword]. In H. Thiesen (ed.), *For flid og god opførsel; Vidnesbyrd fra et eksperiment.* [For Diligence and Good Behaviour: Testimony from an Experiment]. Nuuk, Greenland: Milik.

CBC (Canadian Broadcasting Corporation) (2022). The Current, with Matt Galloway: Survivors are fighting for justice over residential school-like programs in Greenland. *CBC,* February 14th.

Charles, M. & Rah, S.-C. (2019). *Unsettling Truths: The Ongoing, Dehumanizing Legacy of the Doctrine of Discovery.* Downers Grove, IL: InterVarsity Press.

Chrisjohn, R.D. & Young, S.L., with Maraun, M. (2006). *The Circle Game: Shadows and Substance in the Indian Residential School Experience in Canada.* Penticton, BC: Theytus Books.

Corntassel, J. & Holder, C. (2008). Who's sorry now? Government apologies, truth commissions, and Indigenous self-determination in Australia, Canada, Guatemala, and Peru. *Human Rights Review,* 9(4): 465–489.

Diamond, J. (2006). *Collapse: How Societies Choose to Fail of Succeed*. Harmondsworth: Penguin Books.

Ehrlich, G. (2003). *This Cold Heaven: Seven Seasons in Greenland*. London: Fourth Estate.

El-Shabazz, M. (1964). *Television interview*. Available on-line: https://www.youtube.com/watch?v=XiSiHRNQIQo [Accessed March, 2022].

Fleischer, J. (1996). *Forvandlingens år: Grønland fra koloni to landsdel*. ['The Year of Transformation: Greenland from Colony to County']. Nuuk: Atuakkiorfik.

France-Presse (2022). Denmark PM says sorry to Greenland Inuit taken for 'heartless' social experiment. *Guardian*, March 10th.

Friedberg, L. (Director). (2010). Eksperimentet. [The Experiment]. *Nimbus Film*.

Goffman, E. (1961). *Asylums: Essays on the Social Situation of Mental Patients and Other Inmates*. London: Penguin.

Gould, S. J. (1997). *The Mismeasure of Man* (Revised ed.). Harmondsworth: Penguin.

Grønnow, B. (2017). *The Frozen Saqqaq Sites of Disko Bay, West Greenland: Qeqertasussuk and Qajaa (2400–900 BC)*. Copenhagen: Museum Tusculanum Press.

Hay, T.; Blackstock, C. & Kirlew, M. (2020). Dr. Peter Bryce (1853–1932): Whistle-blower on residential schools. *Canadian Medical Association Journal*, 192(9): E223–E224.

Hayner, P. (2010). *Unspeakable Truths: Transitional Justice and the Challenge of Truth Commissions*. London: Routledge.

Heinrich, J. (2019). For a Greenlandic independence. In N. Greymorning (ed.), *Being Indigenous: Perspectives on Activism, Culture, Language and Identity*. London: Routledge, pp. 113–125.

Hersher, R. (2016). The Arctic suicides: It's not the dark that kills you. *National Public Radio*, April 21st.

Hicks, J. (2007). The social determinants of elevated rates of suicide among Inuit youth. *Indigenous Affairs*, 4/07: 30–37.

Higginbottom, P. (2012). *The Workhouse Encyclopedia*. Cheltenham: The History Press.

Horst, R.H. (2020). *A History of Indigenous Latin America: Aymara to Zapatistas*. London: Routledge.

Jacobsen, S. (2015). Greenland commission will probe Danish colonial abuses. *Reuters*, May 2nd.

Jensen, E.L.; Nexø, S.A. & Thorleifsen, D. (2020). *Historisk udredning om de 22 grønlandske børn, der blev sendt til Danmark i 1951*. [Historical report on the 22 Greenlandic children sent to Denmark in 1951]. Available on-line: https://www.stm.dk/media/10146/historiskudredning_final.pdf

Knakkergaard, C.L. (2012). *Et godt liv – trods alt! Tvangsflyttet fra Grønland til Danmark*. [A good life – after all! Forced relocation from Greenland to Denmark.] Havndal: JB Historie.

Kuokkanen, R. (2017). 'To See What State We Are In': First years of the Greenland self-government act and the pursuit of Inuit sovereignty. *Ethnopolitics*, 16(2): 179–195.

Larsen, C.V.L.; Curtis, T. & Bjerregaard, P. (2013). Gambling behavior and problem gambling reflecting social transition and traumatic childhood events among Greenland Inuit: A cross-sectional study in a large indigenous population undergoing rapid change. *Journal of Gambling Studies*, 29(4): 733–748.

Lemkin, R. (1944). *Axis Rule in Occupied Europe*. Washington, DC: Carnegie Institution.

Lynch, J.J. & Minton, S.J. (2016). Peer abuse and its contexts in industrial schools in Ireland. *Journal of Aggression, Conflict and Peace Research*, 8(2): 76–85.

Matthiessen, P. (2021). Vred på Astrid Krag – hun er en kold kvinde. [Beware of Astrid Krag – She's a cold woman]. *Ekstra Bladet*, December 27th.

Minde, H. (2005). Assimilation of the Sámi – Implementation and consequences. *Journal of Indigenous Peoples' Rights*, 3: 1–33.

Minton, S.J. (ed) (2020a). *Residential School Systems and Indigenous Peoples: From Genocide via Education to the Possibilities for Truth, Restitution, Reconciliation and Reclamation*. London: Routledge.

Minton, S.J. (2020b). Setting the scene. In S.J. Minton (ed), *Residential School Systems and Indigenous Peoples: From Genocide via Education to the Possibilities for Truth, Restitution, Reconciliation and Reclamation*. London: Routledge, pp. 1–23.

Minton, S.J. (2020c). Some theoretical touchstones. In S.J. Minton (ed), *Residential School Systems and Indigenous Peoples: From Genocide via Education to the Possibilities for Truth, Restitution, Reconciliation and Reclamation*. London: Routledge, pp. 24–47.

Minton, S.J. & Lile, H.S. (2019). Considering a truth commission in Norway with respect to the past forcible assimilation of the Sámi people. In S. Wilson, A.V. Been & DuPré (eds), *Research and Reconciliation: Unsettling Ways of Knowing through Indigenous Relationships*. Toronto, ON: Canadian Scholars.

Minton, S.J. & Thiesen, H. (2020). Greenland. In S.J. Minton (ed), *Residential School Systems and Indigenous Peoples: From Genocide via Education to the Possibilities for Truth, Restitution, Reconciliation and Reclamation*. London: Routledge, pp. 95–112.

Mørck, A.H. (2020). Derfor gik eksperimentet med de grønlandske børn galt. [How the experiment with Greenlandic children went wrong.] *TV2 Nyheder*, December 8th.

National Indian Law Library (2022). *Meriam Report: The Problem of Indian Administration (1928)*. Available online: http://www.narf.org/nill/resources/meriam.html [Accessed March, 2022].

National Native American Boarding School Healing Coalition (2021). *Coalition Urges Support for Bill Establishing a Truth and Healing Commission on Indian Boarding School Policies in the U.S.* Available on-line: https://boardingschoolhealing.org/coalition-urges-support-for-bill-establishing-a-truth-and-healing-commission-on-indian-boarding-school-policies-in-the-u-s/ [Accessed March, 2022].

Norman-Hill, R. (2019). Australia's native residential schools. In S.J. Minton (ed), *Residential School Systems and Indigenous Peoples: From Genocide via Education to the Possibilities for Truth, Restitution, Reconciliation and Reclamation*. London: Routledge, pp. 66–94.

O'Sullivan, E. and O'Donnell, I. (2007). Coercive confinement in the Republic of Ireland: The waning of a culture of control. *Punishment and Society*, 9(1): 27–48.

Otzen, E. (2015). The children taken from home for a social experiment. *BBC News Magazine*, June 10th.

Pengelly, M. (2019). Trump confirms he is considering attempt to buy Greenland. *The Guardian*, August 19th.

Petersen, R. (1995). Colonialism as seen from a former colonised area. *Arctic Anthropology*, 32(2): 118–126.

Poulsen, R.W. (2021). Greenlanders shipped to Denmark as children seek compensation. *Al Jazeera News*, December 21st.

Rohrer, F. (2006). What's a little debt between friends? *BBC News Magazine*, May 10th.

Rasmusson, D. (2019). Barn fördes bort – experiment utreds. [Children taken away – experiment investigated]. *Sveriges Radio*, March 5th.

Ritzaus Bureau (2022). 'Hjerteløst': Statsminister Mette Fredriksen undskylte onsdag til de seks nulevende af de 22 såkaldte eksperimentbørn. ['Heartless': Prime Minister Mette Fredriksen apologised to the six survivors of the so-called experiment children on Wednesday]. *B.T. København*, March 10th.

Rud, S. (2017). Toward a postcolonial Greenland: Culture, identity, and colonial legacy. In S. Rud (ed), *Colonialism in Greenland: Tradition, Governance and Legacy*. London: Palgrave Macmillan, pp. 119–144.

Sale, R. & Potapov, E. (2010). *The Scramble for the Arctic: Ownership, Exploitation and Conflict in the Far North*. London: Francis Lincoln Ltd.

Seiding, I. (2011). Colonial categories of rule – mixed marriages and families in Greenland around 1800. *Kontur*, 22: 56–65.

Sigurðsson, J.V. (2008). *Det norrøne samfunnet: vikingen, kongen, erkebiskopen og bonden*. [The Norse Society: The Viking, the King, the Archbishop and the Peasant]. Oslo: Pax Forlag.

Skydsbjerg, H. & Turnowsky, W. (2016). Massivt flertal for selvstændighed. [Massive majority for independence]. *Sermitsiaq*, December 1st.

Statistics Greenland (2018). *Greenland in Figures*. Nuuk: Statistics Greenland.

Stannard, D.E. (1992). *American Holocaust: The Conquest of the New World*. Oxford: Oxford University Press.

Therkildsen, J.; Olsen, D.K.; Mathiassen, I.; Petrussen, Î. & Williamson, K.J. (2017). *Vi forstår fortiden. Vi tager ansvar for nutiden. Vi arbejder sammen for en bedre fremtid: Endelig betænkning af Grønlands Forsoningskommission*. [We understand the past. We take responsibility for the present. We are working together for a better future: Final report of the Greenland Reconciliation Commission]. Nuuk: Grønlands Forsoningskommission.

Truth and Reconciliation Commission of Canada (2015). *Honouring the Truth, Reconciling for the Future: Summary of the Final Report of the Truth and Reconciliation Commission of Canada*. Available on-line: http://www.trc.ca [Accessed March, 2022].

Vaaben, L. & Borberg, T. (2022). I 70 år har Eva Illum ventet på et bestemt ord. I går sagde statsministeren det. [For 70 years, Eva Illum has been waiting for a certain word. Yesterday, the Prime Minister said it]. *Politiken*, March 10th.

Watts, C. P. (2011). The 'Wind of Change': British Decolonisation in Africa, 1957–1965. *History Review*, 71: 12–17.

Weichert, R. (2021). Sie wurden für Erziehungsexperimente ihren Familien entrissen. Jetzt kämpfen die 'Experimentkinder' von Grönland für Gerechtigkeit. [They were snatched from their families for educational experiments. Now the 'experimental children' of Greenland are fighting for justice]. *Stern*, December 28th.

Wesley-Esquimaux, C.C. (2007). The intergenerational transmission of historic trauma and grief. *Indigenous Affairs*, 4/07: 6–11.

World Health Organisation (2011). *Suicide Rates per 100,000 by Country, Year and Sex*. Geneva: World Health Organisation.

World Health Organisation (2018). *Suicide rate estimates, age-standardised estimates by country*. Available on-line: http://apps.who.int/gho/data/node.main.MHSUICIDEASDR?lang=en [Accessed March, 2022].

World Population Review (2021). *Greenland population 2021*. Available on-line: http://worldpopulationreview.com/countries/greenland-population/ [Accessed March, 2022].

Notes

1 Since 2009, Western Greenlandic (*Kalaalisutt*, traditionally spoken by the Kalaallit people of western Greenland) has been the sole official language of Greenland. However, the Inuit people of northern Greenland speak *Inukun* (ca. 1,000 speakers), and the Tunumiit people of eastern Greenland speak *Tunumiit oraasiat* (ca. 3,000 speakers).

2 There is little doubt in my mind that an Inuit history of the land would look markedly different to the account I have provided of the colonisation of Greenland, which is based on Western historical sources. I intend no disrespect to that Indigenous lens, nor to those who hold, or who have held it; this is due rather to my unfamiliarity with Greenlandic Inuit oral traditions, and my inability as a non-Inuit, non-Indigenous, non-Greenlandic-speaker to access them.

3 Between 1380 and 1814, the kingdoms of Denmark and Norway were in union. In 1814, under the Treaty of Kiel, Norway was ceded to Sweden, and its offshore territorial claims in the Arctic passed to Denmark. Norway achieved independence from Sweden in 1905.

4 Amongst the many examples of colonial abuses and social engineering in Greenland, KGH regulated marriages – that is to say, they restricted so-called inter-marriages (those between Europeans and Inuit people). Seiding (2011) cited the KGH's 'Instruction' of 1774, under which senior staff were not allowed to marry Inuit women (either 'mixed' or 'unmixed'), and 'common' staff were only allowed to marry women of 'mixed' descent, but not 'unmixed' or European women. In 1788, the Governor of North Greenland, Wille, wrote to the Board of Managers to state his opposition to the marriage of 'mixed' women to Inuit hunters, on the grounds that '...the wife of a Greenlander should be able to prepare Boats, build Houses, sew Tents, flense Seals and other Sea Mammals, prepare Skins, work as slaves and more...'. Seiding saw this as reflecting Wille's view of the 'real' Greenlandic woman as slave, '...a description similar to many other contemporary depictions of indigenous, female bodies, in some cases directly connected to racial slavery' (p. 60).

5 Twenty-six Inuit families were evicted from the villages of Pituffik and Uummannaq in order for the United States to expand the defence area around the Thule air base. In 1953, these 116 people were forced to live in tents for six months – at temperatures estimated at between −12.9 and −20.1°C (Minton & Thiesen, 2020) – whilst houses were being constructed for them 81 miles away in the new town of Qanaaq.

6 Economic realities, as well as 'national consciousness', were in the 'wind', too. The amounts of the loans advanced to the United Kingdom by the United States – which took sixty years to repay (Rohrer, 2006) – did not permit the British to remain as masters of the globe.

7 The Danish Constitution also applies to the Faroe Islands, which became a Danish county in 1816, granted home rule in 1948, and like Greenland, remain part of the Kingdom of Denmark.

8 I use this term deliberately and unapologetically. Raphael Lemkin (b. 1900, d. 1959), the Polish lawyer who originated the word 'genocide', described its process as having '...two phases; one, destruction of the national pattern of the oppressed group; the other, the imposition of the national pattern of the oppressor' (1944, p. 79), As it is incontestable that in their respective experiences of colonisation by Europeans, many Indigenous populations have experienced most, if not all, of the five categories of behavior specified in Article II of the United Nations Convention on the Prevention and Punishment of the Crime of Genocide, I contend that the term genocide is correctly applied in these contexts. See Minton (2020c) and Norman-Hill (2020) for further discussions.

9 Contrast the description by Sámi scholars of the government policy of the forcible assimilation of the Sámi in Norway ('...*introduced* in the field of culture, with school as the battlefield and teachers as frontline soldiers' (Niemi, 1997; in Minde, 2005, p. 7) with that of Adams (1995, p. 5) concerning actions in the United States

> As the Iriquois, the Shawnee, and the Arapaho would eventually all discover, the white man's superior technology, hunger for land, and ethnocentrism seemingly know no bounds. The white threat to Indians came in many forms: smallpox, missionaries, Conestoga wagons, barbed wire and smoking locomotives. And *in the end*, it came in the form of schools (emphases mine).

10 I fully acknowledge that these 'touchstones' reflected my own (non-Indigenous) world-view and understandings, and that it was predictable that they would reflect my training and interests in Western disciplines of psychology, sociology and philosophy, and my nascent familiarity with Indigenous scholarship.

11 Chrisjohn and Young, with Maraun (2006) commented that whilst Goffman '… developed his account with no apparent knowledge of [Indian] residential schools… the relevance of his analysis to Indian Residential Schools has not been overlooked' (p. 62).

12 Carpenter, a pioneering penal reformer, anti-slavery activist and educationalist, had been moved by the plight of destitute children and juvenile offenders she had seen in Bristol, England, in the 1840s, and argued that magistrates and judges should send such children to reformatory schools, instead of prisons. She founded her first 'ragged school' in 1846 and her first reformatory in 1852. These measures led to the passing of the *Youthful Offenders Act* (1854) and the *Industrial Schools Act* (1857), thus establishing the network of residential reformatory institutions in the United Kingdom (Lynch & Minton, 2016). The main objective of these institutions was to inculcate in children the habits of '…industry, regularity, self-denial, self-reliance and self-control' (O'Sullivan & O'Donnell, 2007).

13 Before they were abolished in the United Kingdom under the National Assistance Act of 1948, a number of my own family members spent time in workhouses, including my paternal grandparents. As was the case with the residential schooling of Indigenous children (see Adams, 1995; Minton, 2020a), industrial schools in Ireland (see Lynch & Minton, 2016), the workhouse system in the United Kingdom persisted for long after evidence of harm-doing (including institutional abuse) came to light.

14 The second-placed country in this data set was Lithuania, with 36.7/100,000. The suicide rate in Denmark was 8.9/100,000 in 2016 (WHO, 2018).

15 Hicks (2007) recorded that the average suicide rate in Greenland between 1900 and 1930 was 3.0 per 100,000 of the population, and concluded that the only logical explanation for the dramatic increase in suicide rates among Inuit peoples living in different regions of the Arctic (with similar outcomes existing amongst the sexes and age groups, at different time periods) is that a similar "basket" of social determinants has impacted heavily on Inuit societies. These have been referred to in Greenland as 'diseases of modernisation', as '…they tend to increase in traditional societies (such as circumpolar Indigenous peoples) undergoing rapid social changes, with changes in diet, reduction in physical activity, and exposure to new environmental hazards' (Bjerregaard, Kue Young, Dewailly & Ebbesson, 2004, p. 392). Surely significant also are what have been conceptualised by Indigenous authors elsewhere as the intergenerational transmission of historic trauma and grief (see Brave Heart & DeBruyn, 1998; Wesley-Esquimaux, 2007).

16 For example, Corntassel and Holder (2008), noting that Indigenous peoples are disproportionately the target of state violence, as well as neoliberal reforms – and that 'reconciliation' itself is a neoliberal construct – concluded that such commissions '…have not lived up to their potential for transforming inter-group relations when applied in 24 different countries around the world' (p. 469), and that reconciliation processes permit powerful agencies (such as nation states) to effectively place '… rigid material and symbolic limits upon apologies and truth commissions to promote political and legal stability' (p. 465), requiring the victims of colonisation to become '…reconciled to loss, and this is no basis for a sustainable settlement' (pp. 466–467). Furthermore, it is explicitly stated in Article 46 of the UN Declaration on the Rights of Indigenous Peoples (2007) that the right of Indigenous peoples to self-determination, as articulated in Article 3, does not extend to 'secession': 'Nothing in this Declaration may be interpreted as implying any action which would dismember or impair, totally or in part, the territorial integrity or political unity of sovereign and independent States'. Hence, truth and reconciliation processes, as they are currently conceived, do not

(and cannot) deliver, nor under international understandings be compelled to deliver, on decolonisation as it may be manifested in demands for land return, or movements towards full Indigenous sovereignty.

17 This statement was made in a section of the report under the heading, '*Danmark ikke med*' (literally, 'Denmark not with [us]'). The fact that the verb 'to be' in its relevant tense has been omitted means that it was impossible for me (as a non-native tongue reader of Danish) to judge whether the statement should translate into English as 'Denmark *was* not with us', or 'Denmark *is* not with us', nor whether this apparent ambiguity was an intentional gesture.

18 The Greenlandic Commission thereby worked along a necessarily reduced set of aims: 'to raise awareness of the people about Greenland's history and its impact on the present day'; 'an understanding of the past to advance towards a common future'; 'a long-term process of change of society'; 'generally, building relationships'; 'a process of recognising and learning from the past'; and the acknowledgement that '...atonement is both a goal and a process' (see Therkildsen et al., 2017).

19 In 2010, the then-General Secretary of Save the Children Denmark, Mimi Jakobsen, made an apology for the organisation's role, stating in a press release, 'No matter what, on behalf of the organisation, I would like to unreservedly apologise to the Greenlandic men and women who, in a harsh and unhappy way, were forcibly relocated as children'. As this apology was felt to be too indirect and impersonal, official face-to-face apologies were made to some of the survivors in 2015, in which Save the Children Denmark stated that, 'The children should never have been taken away from their families, and Save the Children regrets having participated in the event' (Jensen, Nexø & Thorleifsen, 2020; translations mine).

20 This 93-page report is, by a considerable margin, the most complete version of the events and surrounding circumstances of the 'experiment', and I would warmly and unreservedly recommend it to anyone who can read Danish.

21 There are very clear 'resonances' with Indigenous experiences elsewhere here, which is why I and others have viewed the larger context of the 'experiment' as being that of forcible assimilation of Indigenous peoples via 'education'. To give a recent example, a Canadian radio programme on this case broadcast in February 2022 was entitled, 'Survivors are fighting for justice over residential school-like programs in Greenland'.

22 The reader may recall the attitudes expressed by Thorning-Schmidt, who was the Social Democrats' leader at the time, towards the Greenlandic Truth and Reconciliation Commission.

23 Our colleague Andé Somby has coined the term *den store norske uvitenheten* ('the great Norwegian ignorance') to refer to the deliberately cultivated and long-standing ignorance in Norway concerning the Sámi (see Minton & Lile, 2019). One could equally talk about 'the great Danish ignorance' in the case of Greenland, or indeed 'the great American and Canadian ignorances', etc., amongst settler populations elsewhere.

24 Whereas Helene was one of the sixteen (out of twenty-two children) who were transferred to the orphanage in Nuuk after spending time in Denmark, Carla Lucia Knakkergaard was one of the six children who was adopted by their Danish foster parents.

25 See endnote 20.

26 Currency conversions as of March, 2022.

27 I cannot find the words to express how deeply honoured I was that Helene asked me to be one of the four guests that she was permitted to have accompany her to the event at the National Museum in Copenhagen. Helene's three other guests were her daughters, Laila Hansen and Anja Otten, and her little sister, Asta Søholm.

28 It is, of course, equally true to say that *some* people know that they are doing the wrong thing and carry on doing it anyway. Although the Meriam Report of 1928 (see National Indian Law Library (2022) stated, amongst much else, that '...the provisions for the care of the Indian children in boarding schools are grossly inadequate'

and '...the labour of children as carried on in Indian boarding schools would, it is believed, constitute a violation of child labour laws in most countries', its 'Recommendations for Immediate Action' were not acted upon, and the last Indian boarding schools closed almost seven decades later. Hence, *some* settler governmental officials (such as Lewis Meriam in the United States, and Peter Bryce in Canada) called out the tragedy that was unfolding in these schools at the time of their operation (and in Bryce's case, fought long and hard against it – see Hay, Blackstock & Kirlew, 2020), but were ignored.

For Diligence and Good Behaviour

Testimony from an Experiment

Helene Thiesen

1

GODTHÅB, 1951

The Visit

Victo is drying the dishes, and Mom is washing up. Victo is also learning how to wash the floor. I am running about, playing with my little brother, Hans. The snowflakes have begun to fall. It's nearly spring when a man calls to talk with my mother. It's the pastor, and I wonder why he's here. Mom takes him into the sitting room, and we children are asked to go outside and play. Afterwards, I ask my mother why the pastor called around. 'I don't know', my mother replies. A few days later, he's here again; and this time, he's with a Dane. Again, I'm curious, but I still get no answer. After they came for a third time, my mother looked completely out of it. She squatted down and explained something serious to me. The pastor had said that I am to go on a long trip, to a country far away, in order to learn Danish, along with many other children. The foreign country is called Denmark. Victo stands in the doorway of the living room, listening. Mom explains that the country is nice and that there are tall trees and many flowers there. 'It's like Paradise', she says. I don't know what paradise is, and Victo starts chattering, and says, 'So shouldn't Helene learn how to wash the floor, then?'. I remain silent, thinking, 'I don't want to go that far away. That's further away than it is to Maniitsoq, and it would mean being away from my mother and my siblings'. The next moment, Mom tells me that she wants to make me some new clothes and asks me what colour of dress I want. I've always been fascinated by the oil stains on the road – the ones that have all the colours – so right away, I reply that it should be a dress of all colours. Then Mom sews a dress with wide stripes of different colours, and it helps improve my mood.

As we walk down to the shop we meet, in succession, Agnethe, Albert, Aron, Eva, and Marie, and Mom tells them that they should go to Denmark, too. 'That way, I wouldn't be alone', I think. Much later, I find out that some of my

DOI: 10.4324/9781003241843-3

relatives, who have been made the same offer, have refused to send their children away. The day before I travel, I'm at home, for the last time, taking my bath in a zinc tub on the kitchen floor. A thousand thoughts fly through my seven-year-old head: 'A trip? But why only me? Who's going to bathe me next time, and where? When will I come home to Mom again? What would my father have said about this, if he were still alive? Mom, are you really letting me travel?'

Leaving

The next morning, my brown suitcase is packed. The sun is shining, and it's the best weather as we head down to the harbour. Mom is carrying my suitcase, and Victoria walking next to her, becomes just as quiet. Hans is walking and talking, as he usually does. 'Can't he be quiet, just for a moment?', I think. As we approach the harbour, the other children who are travelling are already there, with their suitcases. 'Look!'. Victo points to the big ship that I'm sailing on. Curiously, I look over at it, and it feels like there's a hole forming in my stomach as I see the *MS Disko*. 'How do I get out there?', I think. The ship doesn't look that big, either. Can it really take us as far away as Mom says it will?

There are a lot of people on the quay. I feel totally overwhelmed, and I hold my Mom's hand even tighter. Mom quietly and earnestly greets Hansine Holm. 'Take good care of Helene', she whispers to Hansine, who will accompany us to Denmark, as an interpreter. I know her well from the town. She is a nice, tall lady, who looks like a Dane. There are several rowing boats below the quay's stone steps, and we are helped into the boats, with our suitcases, four at a time. When it's my turn, it feels completely unreal. We hug goodbye. I have a hard time letting go of my Mom and my big sister. It is difficult to balance as I go down the big, smooth stone steps, although a foreign man, who is trying to calm us, is holding my hand. It is nerve-wracking to step down into the little rowboat, which wobbles as I board it. I'm to sit on the rope seat, next to Agnethe – our little bottoms might just fit there. We are very quiet, and serious; our mothers are standing right next to the railings, and I look intensely at my mine, and wonder why she is letting me go. 'Why, why? Now it's happening, I'm being taken away!' I think. My mother is getting smaller and smaller, and my arms feel so heavy that I can hardly wave; and at the same time, the situation in all its horror hits me. It hurts my chest, and my tears are welling up. If the others had started crying, I would have done, too. My thoughts race through my head: 'They're taking me away from my mother. So stop them, *anaana* (Mom), I don't want to go!' I stare at the place where my mother is standing, and she gets smaller and smaller – in the end, she is just a small dot. In powerless silence, we board the strange ship. On our approach the ship, I could see many other children onboard, hanging their heads over the railing. The children who joined the ship at Sukkertoppen still look terrified. They look seriously at the six of us who got onboard in Godthåb. The children are Carla Lucia and Kristine Olsvig from Qullissat; Ole Fly from Jakobshavn; Little Kristine Geisler and Johan Andersen from Egedesminde; Henrik Raaschou from Holsteinsborg; and the siblings Ane Sofie and Karl Heilmann, and Joel Kågssagssuk Hansen from Sukkertoppen. The children who joined at Godthåb were the twins, Eva and Marie Holm; Agnethe Tittusen, Aron Levisen, Albert Egede, and me.

In the middle of the ship, a ladder is hanging down the side, with a loose rail made of thick rope. We sit on a small ledge below it, before starting to balance our way onto the *MS Disko*. It seems smart to hold tight on the thick rope. The sea below is completely black. Our suitcases are put on the deck. Awestruck, we stand by the railing and look back towards the harbour, where we can see many small people waving as if their lives depended on it. Some of us cry. 'What are we doing here, anyway', I think. As we sail out of the harbour, and into the open sea, we are called together in front of the dining room and the wide, curved staircase. Hansine translates the messages from a crew member about where we can go on the ship. Then we have to find our suitcases and follow a cabin maid, who shows us where to sleep. We walk down the stairs and get to a passageway with many narrow doors. Luckily, I will be rooming with Agnethe and some of the others from Godthåb. There are four bunk beds in the cabin; I get a bunk and put my little suitcase next to it. There is a narrow cupboard for each of us, and we unpack. Then we wait. Sadly, I sit down on my bed, feeling tense, and whilst I am waiting, I take a look around the cabin. There is a small round window, a small desk, two small sinks, and a towel for each of us. In the bunk, there is a small bedside lamp, by the pillow. Suddenly, there's a knock at the door. A cabin maid, Hansine, and the beautiful Mrs Henriette (our neighbour in Godthåb) show us around – so we have two interpreters. I feel happier when I see Mrs Henriette. She greets me, saying, 'Hello, Helene', with a big smile. I'm nodding, courageously.

There are some very nice toilets on board, where you pull on a chain, and what you have produced disappears immediately. In another room, there is something that they call a 'shower'. Water comes from a round thing at the top – it's like rain, they tell us. It looks very nice. 'Dare I use it? Well, not right away, in any case', I think. We meet a nice man in a white uniform, with two rows of black buttons. I think his uniform is beautiful. He is the school principal, and he shows us the store, with its supplies. Never before have I seen so much food at the same time. In the dining room, they are covering the tables, which are bolted to the floor, and have small flaps around the edges, which can be folded upwards, when the sea is rough. This is to prevent our plates and cutlery from falling. The deck is large, and we can run all the way around the dining room, the kitchen, and the smoking room. And what a view it is from the bow! When we hold onto the rail, we can just about peek over the edge and look down at the giant anchor that is hanging on the outside of the ship. The sight churns my stomach. At the top is the wheelhouse, where the captain sits. His is a much finer uniform than the ones we have seen in Godthåb, when the Marines march along Skibshavnsvejen. We close our eyes and feel embarrassed to see such a smart-looking Dane. As it's time to eat, there are white damask cloths on the tables. I can't remember what

FIGURE 3   Leaving the Colony Harbour (May 23 1951)

our first meal was. After eating, we play on the deck. It's strange not to have solid ground underfoot. And then it's time to go to sleep, in a completely foreign bed. When the lights go out, I feel really unhappy. I miss my mother. I cry a lot, and I get comforted, but there are others to be comforted too, so I fall asleep, sobbing.

How Far Can We Sail, Anyway?

After sailing for a few days, we arrive in Frederikshåb, where more children come aboard. It's Emil Jensen and Gabriel Schmidt. Gabriel cries so much that we others get close to tears, too – it's hard to see him so upset. When we get to Ivittuut, we are allowed to disembark and look at the cryolite quarry. It is a huge, deep hole in the rocky ground, with the clearest, turquoise water at the bottom. All around it are wooden houses of many colours. In Julianehåb, more children come aboard. It is Eli Petersen from Julianehåb, and David Pitsivarnarteq from Kulusuk, and Barselaj Danielsen from Scoresbysund, who have flown to Julianehåb from the east coast. Søren Lundegaard and Bodil Mathiassen sailed from Nanortalik to Julianehåb in order to get onboard. As we sail out from Julianehåb, the adults count us, and I can hear them say that there are twenty-two of us. As we sail on, I see that whilst the mountains are not as high in Julianehåb as they are at home in Godthåb, there are lots of icebergs, and it is very cold. That night, the ship rocks a lot; it feels like rolling back and forth in bed. 'How far can we sail, anyway?', I think, and fall asleep, crying.

After breakfast the next day, I go with some of the other girls out onto the deck, in order to see where we are. We go from one side of the ship to the other and all the way around the dining room. The only thing that we can see, day after day, is sky and sea, sky and sea. One day, we spot some small rowing boats, with one man in each of them. The black-haired men are busy fishing; we wave, and the adults explain to us that they are Portuguese fishermen. They have a mother ship that they row back to when they have caught enough cod. We have plenty of time to explore the ship over the following days. It is a red, long, and large ship. In the middle of the ship, high up, is a large yellow chimney, painted black at the top. There are two lifeboats and about fifteen cabins on each side of the ship. One day, we are admiring one of the twins, who is sitting up in the bow, with her legs dangling over the edge. Suddenly, Mrs Henriette taps us on the shoulders. She has her index finger raised to her lips. She creeps quietly up to the bow, and concerned for twin's life, lifts the girl down. We are then called into the dining room, for a good talking-to. It is spelled out to us how dangerous it is to sit on the bow; none of us should do that again. The next game we come up with is hide-and-seek – there are lots of good hiding places on the ship. One day, we can't find the oldest boys, Eli and Ole, and eventually, we give up and ask the adults to help us. When they shout for the boys to come back, the boys stick their heads out from the lifeboats. Then another serious round of reprimands awaits us.

At Cape Farewell, there are large waves. The dining room windows are completely covered by the huge waves pouring over the ship. It is a terrifying sight, and it feels as if we're sinking. 'I hope the ship can keep on going!', I think. If we go outside, we cannot walk normally and fall over very easily. It's good that the dining tables are fixed to the floor and that the flaps can be folded up around the table edge, so that things do not slip onto the floor. When it's foggy, it's difficult to fall asleep. The foghorns sound deeply, repeatedly, sometimes all day long. It is stressful to listen to them. Some of the boys want to try to climbing the rope ladders, but an adult quickly stops them. When we get to the Faroe Islands, we can see the *MS Disko*'s huge anchor being thrown out, with a violent noise, and later on, pulled in again. The wind is getting warmer and hotter the closer we get to Denmark, and by that time we have spent so much time together that I know almost all the names of the twenty-one other children. We have become accustomed to the daily routine on the ship; we know all of the places that we need to be. We girls often run around on the deck, hand-in-hand, or we play tag and shriek with laughter.

On June 7 1951, during breakfast, we are told to pack our suitcases and line them up on the porch, putting our coats over our suitcases. 'Are we really going to land soon?', I wonder. I get a pang of excitement in my belly; we twitter together on the deck, placing our hands above our eyes, in order to shield our eyes from the sun.

'*Nunasiorpugut!*' We are looking for land. The water is shiny, and the sky is golden, as we see the shore. I remember my mother had said that there are mountains there, and when we get closer to the land, I think about the strange curly peaks that these mountains have and the fact that they are green. 'Try to breathe in!' *'Atagu naavisiuk?'* ('Can you smell it, too?'). Ah, how good Denmark smells. Then we see the houses, which are taller than the church tower in Godthåb. There are even more people at the harbour than there were at the one we departed from in Godthåb. They seem like very nice people, and they are beckoning to us. The boys are very eager to wave again. I think, 'But we don't know them'. So I don't wave.

This Must Be What Paradise Looks Like

After we get down the gangway, we're taken over to a long car, with lots of windows, and steps up to it. 'Are we travelling in it?', I ask. I've never ridden in a car before. The car has nice seats; the twins Eva and Marie sit together, and Agnethe and I sit together. *'Takuuk!'* ('Look there!'), we say, one after the other, pointing at the very long, thick sticks, which are green at the top. Somehow, they are standing on the ground, but we can't work out how. We see them in the glow of the lights along the road, that is, black and flat. The sticks are trees, lining the paved roads. We drive and drive. *'Sumut?'* ('Where are we going to?') we ask Martha, a young Greenlandic lady, who is travelling with us. 'We'll be there soon', she replies.

It is completely dark, and we are sitting dozing when we suddenly hear a new sound. We are no longer driving on asphalt, but on a dirt road instead. Then the bus stops, and we have to get off, and even though it's dark, we can see a huge house. Leaves are growing up the walls. We enter into a large hallway and proceed into a large room. There are windows, but no curtains, and eleven bunk beds in the middle of the floor. The biggest girls, and most of the boys, take the bunks, and I get one. The adults will sleep in some rooms upstairs. We are shown the bathroom and the toilets. There are washbasins in the middle of the floor – so many that we could almost all wash at the same time – and white, glossy tiles on the walls. We get a towel and a washcloth each. When we are all finally under the bedclothes, I start feeling sad again. 'How long will we be here for?', I think. We can hear the wind, and – what's that? It sounds like there's someone hunting. There are gunshots; some of the big boys say that there are Germans out in the woods. There has been a war in Denmark. 'There are probably still some soldiers left', say the boys. 'What are "Germans", and what is "war"?', I think, although I can understand from their tones that these are not nice things. Some of us start to cry. I crawl down in my bed, and when my head is under the covers, I whisper, 'Mom, Mom, come and get me'. That's how I fell asleep. The next morning, we go outside after breakfast. There are the most beautifully scented flowers, of all kinds of colours. We walk very carefully on the big green rug, in our bare feet. But it's not a rug; it's cool, and it springs up under your toes. The adults call it 'grass'. I think that my mother is right – it's beautiful here. This must be what Paradise looks like.

We can see the long, thick sticks now; they are much thicker than those at the Qooqqut fjord. The trees there are a little taller than my mother, but here they reach all the way to heaven. There are green leaves at the top. The thick sticks feel rough and are uncomfortable to touch. Although we push with all our strength, they remain in the ground. We can also see the leaves on the house now. They go all the way up to the roof, and all around the windows and doors, so it looks like a huge peat house with a roof, but underneath, there are bricks. There are strange and violent insects, with yellow stripes, flying around the flowers, and there are nasty spiders, much bigger than those at home in Greenland. A pink thing is digging itself into the ground, and one of the boys picks it up. We scream, and an adult comes along: 'It's an earthworm', she says. The huge lawn is surrounded by forest, and there are also some small, crooked trees on the lawn. We might try to climb them, although this looks difficult.

Before lunch, we learn our first Danish song, 'And we pull the drawer out, and we push the drawer in, whilst the smoke rises up like this…'. Then we go out for a walk. A truck is filled with buckets, spades and packed lunches. It's a bit dark in the woods, but it smells nice. Suddenly, things get very bright. 'It's the big beach at Feddet', the adults tell us. We can just throw off our clothes and go into the water. I think, 'If the water is as cold as it is in Qooqqut, then I won't like it. In Qooqqut we keep our rubber boots on'. I'd rather sit down and dig my feet into the sand, so that nobody can see my ugly big toes. When someone approaches, I also hide my ugly thumbs. We have collected snails in our buckets, and when we get back, and go to bed, we put the snails in the bathroom. The next morning, when we go into the bathroom, our eyes nearly pop out of our heads. The snails have been crawling up and down the tiles. Their trails look nice, but we are told to catch them and put them out in the garden.

Big Kristine always plays with the big boys. She climbs high up into the trees, just like them, but one day she falls. She screams frantically as we run over to see what has happened. She has broken her arm, and a doctor comes out to her, in a black car. When she comes back, her arm is in a plaster cast.

Beneath the windows of Fedgården, as this place is called, there is a large earth bank, where flowers grow. One day, I watch the handsome old gardener working. He picks a completely round and red flower, stands there admiring it, and asks if I want to taste it. I stare at him as if he was mad. Just as I am saying to him, 'No, I don't eat flowers', the gardener puts it into his mouth. 'He's eating the red flower!' I cry. It's a little tomato. The gardener is also a janitor, and later on, he said that the little Greenlanders were easy to deal with. One day, we have guests, so some of us have to dress up. Something similar to a Greenlandic national costume has been made for Ane Sofie, and her hair has been set up in a *qilerteq*, a bun. She has to sing *'Piitaq Aasiannguaq'* ('Little Peter Spider'), and we help by singing along with her, although I think that it's silly.

We hang around with the adults whenever we can. One day, a couple of us kids go along with Martha as she bathes. The bathroom is very nice; they have a large white bathtub. We hang off the edge of the bathtub and look at Martha, who has a small mirror in the bathtub and a funny little thing in her hand. She is plucking her eyebrows. 'Oh, that must hurt!'. She also plucks the black hair on her legs. 'That's why she's so beautiful', I think. I am not allowed in the tub, because I have bad eczema in my elbows and the backs of my knees, which are covered with a thick layer of white ointment.

One day, there's hunting at Fedgården, and we are interested in what they catch in this country. The hunters have changed into smart clothes – they all wear similar caps, jackets, plus-fours, and rubber boots – and there is a horn to ride to. We get the hunter with the horn to ride in front of us, and he blows it. It must be to scare the animals away. On another day, they are dragging a giant animal along. It has four short legs and is a strange pink colour, has a curly tail, and a very thin layer of short, stiff hair on its body; it smells, and it's very fat. The men struggle with the animal, which is squealing so loudly that we have to put our hands over our ears. It is hung up by its back legs, and a man puts a big knife into its body, which makes it squeal even louder. We girls run away, screaming; later on, we get pork for dinner.

We are learning a song –'In a forest, there was a cabin, with a hunter at the window' – and the hand-actions (e.g. we have to mime a house and a hunter shooting). Afterwards, they tell us that one of the days, we will visit the Queen. There is bustle everywhere; even the grass gets an extra mowing. It seems that she's very important, because we all need to dress in our nicest clothes; the boys are wearing their white anoraks, and we girls wear nice dresses with white collars. Ane Sofie puts on her Greenlandic national costume, and we wait and wait, behind the railings. It's really boring. 'What are we doing here? Will I be able to go back home to Mom when she's visited?', I think. Just then, a big black car comes along. It does a lap of the courtyard, then a man in uniform jumps out, and opens the door for the Queen. She is wearing a narrow skirt, with a matching, low-waist jacket, and a small hat with a bow at the front, in the same material as the jacket. There's a net over her face, and she has gold earrings and a gold bracelet. She is beautiful and has wavy hair. This is what a queen looks like – a fine lady. Mikisoq hands her a bouquet.

2

HISTORICAL BACKGROUND

New Times Coming

For centuries, Greenland has been a Danish colony. In 1925, a new law was passed, the long-term aim of which was to open up Greenland to the outside world. To do this would require advanced development. Plans were implemented in each district of the colony; and in the largest towns, secondary schools opened, and improvements were made in the teaching of Danish. Back then, my dad was seven years old. Just after the war ended in 1945, when American naval vessels often came to the Greenlandic towns, thus connecting them to the outside world, a delegation from Greenland sailed to Denmark to discuss the need for, and expectations of, new times in Greenland. The delegation suggested that in order to bring Greenland forward, there should be improvements in the education system; however, the time was not yet ripe for a big leap into a new scheme. These measures were postponed and received heavy criticism, especially from the officials who were isolated in Greenland during the war. In 1948, the Danish Prime Minister Hans Hedtoft Hansen met with the two national councils, who assembled in Godthåb. In those days, the councils could only meet once a year. Hans Hedtoft asked the assembled councils directly whether Greenlanders themselves wanted the country to open up and to develop into a modern society, with all the consequences that such a move would have. A few years previously, Danish statutes and laws began to apply in Greenland. These included, amongst other things, regulations on hunting. At that time, and over the heads of the prisoners,[1] the colonial leadership determined what could be caught during the various seasons. Back then, the prisoners lived solely from what they could catch. A prisoner who had shot two reindeer outside the new conservation regulations would be fined – please pay fifteen kroner![2] Such a person might never have owned so much as a penny, yet his very existence was threatened by these new laws.

DOI: 10.4324/9781003241843-4

In March 1950, a package of laws on dealing with Greenland was presented to the Danish parliament. One of the new laws was the School Act, where the school system was separated from the church, and the teaching of Danish was to be strengthened. Greenlandic was then the language of instruction in schools. In the media at the time, Greenlanders debated whether Greenland should be Danish-speaking. The person in the debate who was the most eager to make Danish the main language was the Greenlandic politician, Augo Lynge. He wanted the Danish language to be used all the way through school, from kindergartens to primary schools, and in the secondary and high schools. His vision for the future was one of all Greenlanders becoming 'good Danish citizens', and he was so occupied with this that he pushed ahead with his ideas. Lynge's own children attended the 'new Danish school' in Godthåb – one of his daughters, Helga, was in my class.

After World War II, a medical doctor, Dr Ludvigsen, returned to Greenland after having been away for ten years. It disturbed him to see that the poverty and misery had worsened during the war years. Greenland was in the middle of a transitional period, a development that the population was just watching. Many of the Danish emigrants found it difficult to believe that the Greenlanders would be able to participate in the construction of Greenland at all, and they expressed their sense of hopelessness in helping the Greenlanders. Dr Ludvigsen highlighted the first private orphanage in Lichtenau in southern Greenland as a model. There, the seven children could speak and understand Danish; they were healthy, well-behaved, and well-dressed. He expressed his appreciation of the help that the Danish Red Cross had provided, and this is where he got the idea of creating more orphanages, in order to improve the health of Greenlandic children. After reading his recommendations, the Danish Red Cross decided to build an orphanage in Godthåb.

In 1949, Save the Children applied to send Greenlandic children to foster families in Denmark. The application was rejected, but Save the Children did not let the idea go. Contact was established between Eske Brun, the Head of the Department of State for Greenland, and Save the Children. Shortly after, on July 20, 1950, the Greenlandic Council met, and the Chairman, Governor P.H. Lundsteen, presented Save the Children with the proposal that about twenty children could be sent to Denmark. First, during the summer, the children would stay at one of Save the Children's holiday camps, and in the winter, they would stay with Danish foster families. First and foremost, the children should be orphans. The chairman's proposal was adopted by the national councils, after a forty-eight-hour period of reflection. There were eighteen votes in favour of the proposal, and four against. At the same meeting, a proposal made by the Danish Red Cross to establish orphanages in Greenland was also on the agenda; everyone voted in favour of this. There was no mention in the minutes of the meeting that the twenty children who would be sent to Denmark would move into a Danish Red Cross orphanage on their return. This important information for the families involved first came to light six months after we, the twenty-two children, had been sent to Denmark.

On December 1, 1950, the Greenland school principal was asked to take steps in selecting a group of twenty-two children which, as far as was possible, should be equally distributed by gender. The Ministry for Greenland in Denmark wrote:

> Since it is thought that after a one-year stay in Denmark, this group of children will return to Greenland, and be admitted to the Danish Red Cross's orphanage which will hopefully be established by that time, and will continue their education in the upcoming bilingual schools, emphasis should be placed on selecting children with the highest possible intelligence quotients.

Priests, catechists, and assistant teachers from Qullissat, Jakobshavn, Egedesminde, Holsteinsborg, Sukkertoppen, Godthåb, Frederikshåb, Julianehåb, and Nanortalik on Greenland's west coast, and from Ammassalik and Scoresbysund on the east coast, all spoke in favour of the new assignment and started to find suitable children for this experiment. However, at that time, many children had not yet mastered the Danish language. From the answers telegraphed back to Copenhagen, it was clear that the intentions for the 'experiment' were not understood; finding enough children would be a difficult task. On April 1, 1951, the governor wrote, 'No subjects have been notified'.

The sending of a small group of children to Denmark, to form the core of the upcoming school system, became a prestigious project. It became a matter of urgency, with rapid communication, in the form of telegrams, flowing back and forth between all the towns involved. The governor suggested that the age limit should be lowered to five years and that the high intelligence criterion should be dropped. The Ministry for Greenland approved both the reduction of the age limit and the dropping of the high intelligence requirement. Furthermore, as a sufficient number of orphans did not exist, the 'experiment' was extended to include motherless children as well. Shortly thereafter, the school principal was able to write that twenty-two suitable children had been found. Of these, fifteen were motherless; however, that did not mean they had no families at all, because most of them had fathers, older siblings, and other close relatives nearby. When the children of Godthåb were to be found, as the last six, the original plan again went awry. We all had mothers; yet we were sent to Denmark. Back then, I was seven years old.

My Family

My father was born in the colony of Godthåb in Greenland on December 21, 1918. His name was Hendrik Jørgen Karl Samuel Kristoffersen, and he was a telegraphist. My grandfather was a prisoner named Aron Martin Kristoffersen, and my grandmother was called Bibiane. My father and my grandfather grew up in a small, white house on top of a hill in Islandsdalen, Godthåb. To the west, there was a beautiful view over the Kookøerne islands; to the north, a view of the northern country, and to the east, a view of Sermitsiaq, Godthåb's famous mountain. To the south were Hjortetakken and Store Malene, which peaks at 759 metres; Lille Malene is 420 metres in height. At Grandpa's house, there was a small hallway, and a large room that served as kitchen, living room, and bedroom – that's how I remember it. When we visited him, we went right inside from the door, and on the left, grandfather lay on his wooden bed, underneath a huge quilt. He had a small nightstand, with a coffee cup and a packet of sugar candy on top. Under the bed, there was an enamel pot; I was surprised that Grandpa used it. He was always happy to see us and said, '*Iggu!*' ('How sweet you are!'), and then gave us a piece of candy.

I was told that my father had a brother, named Hans, who died from tuber-culosis at a very young age. Another brother was Ignatius, who had moved to Egedesminde as an adult, where he was a teacher and, later on, a politician. A third brother was Jonathan, who lived in his grandfather's house in Islandsdalen until his death. He married very late in life, and they had a little boy who was baptised Hans. My grandfather's brother, Kristian, lived next door – he was married and had several daughters.

My mother's name was Magdalene Bilha Augustine Rasmussen. She was born 1922 in the Sukkertoppen colony. Her father was a prisoner named Michael Hans Karl Rasmussen, who was born in 1896, and her mother was Marie Helene Rosine (née Heilmann), who was born in 1898. My maternal great-great-grandmother had been a sort of a foundling. The winter when she was born had been terribly hard. Even the sea had frozen solid, which was very unusual in Sukkertoppen. The prisoners could not hunt, and the hunger and distress spread; many people died of starvation. When a thunderstorm broke, the people could finally move between the different settlements and then the supplies could arrive again. A family who were travelling at the time passed the village of Napasoq, near Sukkertoppen, and they were greatly upset to see that the entire population of the village had been wiped out. They spotted a little baby, who was trying to feed from her dead mother's breast. At first, they had thought the baby was dead, but they suddenly heard a soft moan. The baby survived, although she was severely crippled. That baby was my great-great-grandmother, Karoline Sabine Agathe Lynge, who was born in 1845.

My mother was the oldest of five children. One of my aunts was named Sofi-aaraq Holm, and the other was Karoline Heilmann. I never got to know my mother's brothers. One of them, Vittus, has drowned in Thule when he was very young, and the other was named Michael. He sent my grandmother to live with another man after my grandfather died in a kayaking accident; my Aunt Karoline told me that. Their maternal grandfather was Ivar Hans Diderik Heilmann (known as Tidalik), who was born in 1873. Their paternal grandfather was Vittus Klaus Vilhem Rasmussen, who was born 1859; he was a small man, with a loud, clear voice. He was a ship's captain and the postmaster at the time. In his day, a postal kayak trip from Sukkerrtoppen to Holsteinsborg cost four pounds of bread, one-eighth of a pot of brandy, half a pound of English tobacco, and eight rigsdaler [national 'dollars']. An instruction from 1873 read, 'Regular mail is sent three times a year: In winter or spring, in summer and in autumn, departure and pick-up times are determined by the inspector'. My grandmother died of pneumonia when my mother was ten years old. My Aunt Karoline was told that she would stay with Sofie and Lars Møller, and my mother and Aunt Sofie would go to the children's sanatorium in Sukkertoppen, where they had to stay until they reached confirmation age. After they were confirmed, they had to move back home to their grandmother. There was no confirmation ceremony for them.

During World War II, Greenland lost its connection to Denmark, and Greenland became self-governing. That was when Godthåb became Greenland's capital, and the administration assembled there. An American and Canadian consulate was established in Godthåb in 1940, and the following year, the Greenland Treaty was signed, which allowed the United States to build military bases in Greenland. So America became Greenland's connection to the outside world. The worst period for supplies was in the summer of 1944. Supplies such as rye flour, dry milk, and margarine simply ran out. At that time, the population of Greenland was 20, 184.

In my mother's home they had blubber lamps, and they used either liquid blubber or cod liver oil as fuel. They cooked the cod liver and sieved it, before using the oil for the lamps. Petroleum first became available in the 1950s. They used brambles and peat to fire up the stove, which the children were supposed to help gather. My mother lived with her family in a small house, just across from the church. It was one of six houses that formed a small enclave in Sukkertoppen which was known as 'Greenland's Venice'. When high tides came, it was as if they lived on a tiny island, because the water rose right up to the six houses. My mother and her siblings went out to work early. They packed catfish in large half- and whole kilo blocks. When they were really busy, they had to start work at half past three in the morning. When a large catch of cod came in, they were employed as child day labourers. Even the Danish children served as day labourers, and everyone helped, my aunt thought; however, she and the other Greenlandic children thought that the Danish children were lucky to be able to use their earnings as pocket money.

Fortunately, by the time my aunt was married, petroleum had arrived. Even after having children, she continued to work in the fish-packing plant with her husband Samson Heilmann. It was hard work. None of my aunt's ten children attended any form of daycare. At nine o'clock in the morning, she ran home from the fish-packing plant, nursed the youngest ones, and then returned to work. When the children were a little older, baby bottles were put into their mouths, and then they had to look after themselves until their mother came home from work. It was a busy little community. The 'Coalman' lived in one of the houses, and a woman named Kammiortoq, who made fireplaces lived in another. In other houses, there were fishermen and hunters, but there were also houses with store employees from the KGH,[3] and catechists.

Illiteracy has been largely unknown in Greenland since the 1850s. In Godthåb, there were several secondary schools, and my father and his brother Ignatius, amongst others, received their education from the college. My father became a telegraphist, and my uncle was catechised and educated. My uncle told me that he and my Dad competed to be the best in their subjects, but my father was better at gymnastics than he was. My father's training as a telegraphist meant that he had to make a work-trip to the colony of Sukkertoppen, where he met my mother at a dance in the assembly house. He fell so much in love with her that he took her back to Godthåb. My mother and father were married on January 3, 1942, in the church in Godthåb. Their wedding was the first wedding broadcast over the airwaves on Greenland Radio's first Greenlandic programming.

On January 5, 1942 you could hear, for the first time: 'Greenland Radio, Godthåb, at 475 medium wave. Here is the radio journal with the latest news'. Then a short broadcast of news from the outside world followed; then, the news from Greenland, with various official announcements, as well as the sailing- and motor-boat positions. Right after the news was broadcast in Danish, the news in Greenlandic was broadcast. This type of programme had to begin from scratch, as such a service had never existed before. My father was, therefore, one of the pioneers; he received the telegrams for the radio journal from Godthåb, which contained the most important news. In addition, a brief summary of the news was sent out every day on the telegraph, which was received and transmitted by all of the radio stations in Greenland. Every Danish family received a small newspaper daily, the text of which filled an A4 page. This system had come into being before the war, when the Greenland Board sent the main newspaper from Copenhagen, but this became impossible during the wartime occupation of Denmark. The main newspaper was also translated into Greenlandic and distributed, but it was only available in outlets in the colonies. Here you could see the local people in the evenings flocking to read the latest news. We had a radio receiver at home. During the war, one could order a radio from the Department for Greenland in New York. In 1941, about 100 households owned a radio; by 1945, this number had risen to 400. During World War II, school supplies from the Red Cross in the United States were also sent, and every household received a free bottle of American red wine.

After the War

My older sister was born on October 26, 1942, in Sukkertoppen. Shortly after giving birth, my mother went to Godthåb to stay with Aunt Sofiaaraq, so that she could help her. Water had to be fetched, and clothes had to be washed. Auntie also got a job; she washed the floors in my Dad's workplace. When it was payday, Auntie also had the task of picking up a banknote from my father in the telegraphy station, and then going all the way to my grandfather's in Islandsdalen, and handing it over to him. On the way out to my grandfather's, Auntie used to pass a big red house, where a handsome young man, who had his eye on her, lived. Auntie was shy, but she didn't mind going out with the money to my grandfather, because Hans Holm was always ready to admire her from a distance. Through a pair of binoculars, he watched her walk up from the Hans Egede statue and reappear by the US Consulate; he also watched her return journey, passing by the KGH building, and back to our house. They were married when she was only eighteen years old, and she had her first child, who died soon after. Auntie and Hans Holm had eight children in all.

Eske Brun was the national leader in Greenland during the time of the post-war liberation of Denmark, and shortly afterwards, the Greenlandic Government was established. In 1948, Greenland was still a colony, and there were two Councils – one in northern Greenland, and the other in the south. Hans Hedtoft had asked these councils – 'What future do you want for Greenland?'. What they wanted was an openness to the outside world and the modernisation of the school system. KGH's commercial monopoly was abolished in 1950. One of my father's old books reads:

> By royal resolution, a Commission was set up in 1949, representing Parliament, the Greenlandic Councils, and the Danish administration. The Commission was supplemented by a significant number of experts. The Commission concluded with a strong report, explaining these issues and proposing future work in all fields. On the basis of this, in 1950, a number of bills were submitted to Parliament, for example, on the positions of the Governor, the Dean and the School Director. In addition, a law was passed on the pursuit of occupations in Greenland, which laid down rules for the right to engage in commercial fishing, fishing and hunting in Greenland.

The pursuit of private occupations was reserved for Danish nationals residing in Greenland, as well as other Danish nationals, with special permission from the Danish Prime Minister. The first private department store opened in 1950. This was Ole's Department Store. Soon after this, other shops opened. Kristian Bure's 1952–1953 yearbook for the Danish Tourist Association states:

> The sailors, geologists, geodesists, the engineers who searched for coal deposits, the fire department, the people who set fires and planted flags, artists, writers, film-makers and photographers, and the legal expeditions all came. Greenland is being built up; a rapid process of institutional establishment is commencing.

I experienced this time – when we Greenlanders were to become 'good Danish citizens'. There was so much construction going on that they had to bring in Danish tradesmen. First, these workers had to build their own accommodation, including the red artisan barracks behind the orphanage, near the cemetery. Fathers in local families, who had been educated and lived in one of the developing towns (Godthåb, Julianehåb, Holsteinsborg, and Jakobshavn), received a subsidy from the colonial government. They were permitted to build their own houses, according to a housing benefit model. When my father was a telegraphist, he built himself a white house right next to his place of work, the telegraphy station (which is known in Godthåb as 'the Duplex'), next to the river and the small footbridge.

My Early Childhood

I was born on April 21, 1944, in the bedroom in our house in Godthåb. My mother told me that my nose was so small that it could hardly be seen. I've never been told what time of day I was born at, nor what I weighed, nor how tall I was. My little brother was born on November 3, 1946. We were a happy family and lived comfortably. In the living room, we had a dining table with six chairs, armchairs, a sofa, and a coffee table with a tablecloth. There was a low, dark brown bookcase with our radio receiver, pictures on the walls, curtains, kerosene lamps, and potted plants on the windowsills. On our book-shelves, my father had many large, black, heavy books: '*The Greenland Council's Negotiations from 1938–47*'; '*Reports on the Greenland Board of Directors, Nos. 1–1949*'; '*Reports from the Medical Expedition issued by the Board of Greenland, 1947–48*' ,and many others. In the winter, it was easy enough to keep the food fresh, because some long nails had been hammered into the side of the house, outside the kitchen window. Razorbills, eider, and grouse were hung there, in nature's own freezer.

I was very jealous of my little brother. He was adored because he was a boy; I could sense that, early on. My dad got a camera when my little brother was born, so there are childhood pictures of Victoria and me, too. One day we were allowed to go over to see what our father was doing at work, and of course, my little brother had to go first. As he was on his way up the cement stairs, I saw my chance to give him a little bit of a push. All of a sudden, he was lying at the foot of the stairs, screaming wildly. 'It's not too bad', I thought, until I noticed a bump on his forehead, and we had to hurry back home. In the kitchen, my mother found the biggest knife that we had in the drawer, and she pressed it against the bump. I shut my eyes for a bit, until it dawned on me that my mother wasn't going to cut the bump off – she was just going to cool it down with the blade. I was very embarrassed by my jealousy.

One day I was out alone playing. It was lovely and cold. The river just outside the door had frozen, and the sun was shining a little. I had a sealskin that I used for sledding. I played by myself, going up and down the little hill outside our house. Then I decided I would try sledding on the other side of the river. It was so wonderful; I went further and further away from the river. Meanwhile, it had become dark. Suddenly, I didn't know where I was, or where our house was. I walked around scared, and I had to pee. Eventually, I couldn't hold it in any longer, and the pee ran hot down my legs. I started to cry, and through my tears,

I could see there was light in some of the windows. My legs got cold, and I heard my mother cry out in the dark, *'Iliina, sumiippit?'* ['Helene, where are you?'] Then I could see her in our house's kitchen doorway, and I ran up to her, still crying. As soon as I got inside, she saw that I had wet my pants. She lifted me up onto the kitchen table, washed me, and gave me some dry pants. Then I was given some hot *suaasat* (soup cooked with razorbills, onions, and porridge rice). It was great to get into the warm. My mother had been breastfeeding my little brother and had fallen asleep afterwards, forgetting that I was still playing outside. I was around three years old at the time.

In order to light the coals in their stoves, in summer my mother and all the other women had to go up into the mountains and gather heather. Mom collected up as much heather as she could carry and tied it together with a cord. Then she wrapped a piece of cotton fabric around her forehead, like a giant headband, to protect her skin from the cord around the heather. The heather would hang down the women's backs, and in this way, they could carry large bundles of heather home, using their arms behind their backs to give their loads additional support. In the old days, the Danes called the women 'heather hunters'. When the stove was lit, it was hard to get the heather smoke out of the house, but it smelled lovely. Mom collected so much heather that we had a store in the basement, with enough for the whole winter. When my mother was going out to gather heather one day, Aunt Karoline was helping her, and we youngsters were being taken care of by one of our neighbours' older children, in the red house at the far end of Skibshavnsvej. Whilst we were there, a tall young man with curly hair came to visit. He called me out into the kitchen, stood in front of the stove, and whispered to me to lie down and take off my underwear. This seemed very strange to me. Just then, we saw my mother and aunt coming home, carrying a bundle of brambles. My aunt came in to see what the young man was doing; she scolded him, chased him away, and helped me to get my underwear back on. The young man went completely red in the face, and he ran out of the house.

We also had a cellar where the coals were stored. Inside the kitchen was a trapdoor with a brass ring in the floor, which led down to our stores of potatoes and dry fish; the coals were at the far end of this cellar. I thought that it was creepy down there. When my mother wanted to go shopping alone, we were shut in there. I used to cry out wildly, but my mother didn't come and help. Maybe that's why I often had nightmares. I used to dream that I had woken up in a dark brown, square grave. I could only see a single crack of light, but I couldn't climb the brown earth sides. I screamed for help, but I had been buried alive in the dark hole. When I had these nightmares as an adult, my husband would wake me up, and comfort me. 'Have you been locked in again?', he would ask, and I would reply that at least I had not been this time. Only when, many years later, I heard my big sister say how scared she had been when we had been shut in down in the deep, dark cellar, it dawned on me. As soon as I heard this, I wept, and after that, my nightmares disappeared.

Branded

Mom, Dad, and my little brother were still asleep when my big sister and I were down in the living room. It was getting bright outside, and I went to the window and looked out at the newly fallen snow, which looked lovely. At first, my sister and I whispered together. 'Get down. You know you're not to stand on the arm-chair', Victo said, pushing me. I jumped down, and started running around. As I got near to the stove, I suddenly felt a nice warm wave against my bare arms. I was curious and asked Victo whether it was hot – the stove looked completely red hot to me. Irritated, she took me by the arm, and led me over towards it, grabbed me by the wrist, and pushed my hand into the stove, saying, 'Want to feel?' It made a sizzling sound; I screamed wildly and started sweating. Through my tears, I could see my crumpled, reddish skin, and to my horror, I could see the tendons in the back of my hand. I screamed even louder as I heard my mother and father coming down the stairs. 'What's going on?', they asked, at the same time. I screamed that it was Victo, all the while pointing at her. They looked at my hand, horrified. My mother rushed me out into the kitchen, quickly poured some water into the enamel washbasin, and put my hand into the cold water. I sweated, screamed, and cried for a long time. Victo looked so guilty, and she got scolded so much, that I was rejoicing even in the midst of all my pain. She looked really angry at me.

A few days after this, my father promised us a new sealskin, so that we could go sledding on the hill just outside the house – he had been given the skin by our grandfather. Between our neighbours' house and ours, there was a rounded slope, leading down to the river. When there was a lot of snow, it made a nice run – we could sled all the way down to the other side of the river. With the hairs of the sealskin facing downwards, and especially when there were two of us on the skin, we could build up a lot of speed. We got snow all over our faces, and we sled again and again, until we had worn all the hair off the skin completely. Then my father decided to make small skis for us. He got empty barrels from the butcher in Godthåb, and he separated the staves. Then he cut straps of sealskin and hammered them onto the edge of the staves, so that we could put our feet in to them – and then we had the first pairs of skis of our lives. At first, it was difficult to balance on them, but we learned, and we spent many happy hours skiing outside the house.

My father's sister often came to visit us. She always said a lot of kind words to me, quickly, and in a muted tone. One day, she asked me to purse my lips, like she was doing, and then a whistle came from her lips. I looked intensely at her lips, and practiced and practiced until a small whistle came from mine. As soon as I succeeded, she took her middle finger and thumb, and 'pinched' the position into my lips. Every time I met her at home, or on Skibshavnsvej, she wanted to hear me whistle, and then she'd whistle, too. After a while, I didn't think that whistling was much fun.

In the winter, mother washed clothes in a sink in the kitchen. The clothes were hung up to dry on a homemade wooden drying rack that hung above the tiled stove. In the summer, she would wash clothes in the river that ran right outside our door. I followed her, as I was interested, and I was impressed with how quickly she rubbed the clothes clean on the washboard. It looked difficult when she had to twist a sheet. There was a large nail shaped like a hook on the wall of the house in front of the kitchen door. From there, a cord went out to a pole that stood in the ground. The clothes hung there and fluttered in the wind.

While playing outside one summer day, we suddenly heard a strange sound. We headed to the kitchen door to see what it could be, and there was a huge four-legged animal with a lot of hair on its head. We fled in terror to Mom, who told us that it was a horse. For a long time, we did not dare to go out playing, but when we found that the horse was tied up, we went out playing again. It turned out that the water supply, which came from several wells, was drawn by horse-and-cart. They used three horses to handle this task, and they grazed just outside our house.

Once, I walked a little bit too far away from our house, and I suddenly saw a short lady. She wasn't much taller than me and my big sister, but her head was much bigger than ours, and she looked old. I was terrified of her, because I thought that she looked completely wrong. A little later, another lady came by, who looked just like her, and I ran home, frightened. Mom told me they were twin dwarves. When I had to walk past their house, I went around in a long arc, even though I had to walk out onto the wet marsh.

One day, my father was hammering and knocking, by the door in front of the bedroom. Curiosity drove me up there. 'What are you doing?' I asked, and he told me that he made a toilet. We were used to sitting on the potty, and Mom and Dad sat on a tin bucket, but now Dad had made a hole in the middle of the floorboards and had fixed a hinge at the back. The toilet paper was placed to the right, and he fixed a bucket just below the hole. So we had our first toilet. It made a 'plumpf' noise when there wasn't so much in the bucket, but it wasn't too nice when it got full.

My mother collected lots of tall bottles, and she poured their contents into a huge tub, which stood in the hallway for several days. One day, she poured the contents of the tub back into the bottles, although she didn't re-seal the bottles right away. My big sister and my mother were going down to the KGH; I had to look after Hans, and we were told not to touch the bottles. Whilst they were gone, I thought that we should taste what was in the bottles. If we only took a small sip from each bottle, Mom wouldn't be able to see what we'd done. It didn't taste good; and then we started to feel dizzy. When Mom and Victo came home, Hans and I were wandering around, singing like crazy. They couldn't help but laugh at us, drunk for the first time in our lives, because what we had tasted was *immiaq*, homemade beer. Hans and I went to bed, and we slept really well that night. It was fun to watch Dad open a bottle of *immiaq* when we had guests, because a lot of foam poured out of the bottle.

On Sundays, when the weather was good, we went hiking. In front of the hospital, there was a nice beach, mostly of flat stones. It was fun to tease the sea, shouting, 'The waves can't get me!', and running after the waves as they receded, and then running as fast as we could as they broke back onto the beach. We did this again and again, until our shoes were soaked. It was also exciting to be down at the harbour and watching the waves billowing against the bulwark. We would stick our heads through the railings and look down at the waves. One day, my little brother pushed out too far, slid right between the railings, and plunged into the waves. I saw his head disappear under the water, but fortunately, it reappeared quickly. I looked on, speechlessly, as the adults all ran out to rescue him. Fortunately, there was a rowing boat nearby. 'Hold on!', they kept yelling to him. Occasionally, he would lose his grip on the bulwark and slide under again, but fortunately, a man caught hold of him and lifted him out. What a horror – and then there was the long walk home. We had to walk past Hans Egede's old residence, up the hill, along the river and past the waterfall before he could put on dry clothes again. We kids looked out for ourselves as best we could.

The Woman in the Snowdrift

There was a large brass bell on the end wall of the bailiff's office. It was mounted high up, and a rope was tied to the bell, in order to ring it. Every day, men gathered around it, and when the bell rang, they got up and went off to their work at the colony port. Some unloaded sacks from ships, others shovelled coal from a huge pile, and still others either worked in the warehouse or went over to the fish-packing plant. Some men were driven to the docks, on the city's truck, to work in the warehouses over there. There was a lot of hustle and bustle.

All that we could do was to stand and stare when we were visiting Aunt Sofie, Uncle Hans and his parents. All around their house there was angelica, potatoes and poppies, enclosed by a white painted fence. There were great views from their house – the big red wooden church was at the bottom of their hill, and behind that was the Danish cemetery; and there was a bronze statue of Hans Egede on top of the nearest mountain. My Uncle Hans would look out over the sea, examine the sky, and then predict the weather; he would know then if he would be able to go out fishing the next day. My uncle stayed at home with his parents for the whole of his life, even after he got married. His younger brother also lived with his parents for many years.

One summer day, while we were sitting in front of our house playing with soil and old tins, and we heard a strange humming in the sky. Victo and I screamed, and called out to mother, shouting, 'Help! There's a monster in the sky!'. Our hearts hammered with terror, and we clung onto our mother's skirts. She wriggled free, so that she could stick her head out of the kitchen door. She saw that it was a 'Catalina' – a type of aircraft that could land on water, which was used in Greenland at the time. She reassured us, explaining that it was a type of machine that could fly up into the air. But we stayed inside for the rest of the day anyway.

Every two weeks, large, white parachutes were thrown down from the American aeroplanes; this happened from 1941 up until the mid-1950s. These held mail, and goods, for the Americans and Canadians. The only man who could communicate over the radio with the American aircraft before they dropped the goods was named Winstedt, who owned the first radio shop in Greenland. Later on, when we got a little braver, when we heard the American aeroplanes we would run over to Store Slette, Narsarsuaq, which was a bog where they also dropped down sweets to the kids.

In the 1940s, the town of Godthåb was not large; everyone knew each other, who was related to whom, and who lived in all the various houses. When the weather was bad, my cousin Ane-Johanne and her siblings would get bored, and their mother told them to sit in front of the living room window, count the houses in Godthåb, and try to remember the names of the people who lived in them. So they did. They started from Islandsdalen and finished up at the Greenland cemetery. There were about 500 houses, and they knew who lived in them all – children, adults, and grandparents. Starting from the north, the first area was Myggedalen, which was so-named by the Danes, because there were so many mosquitoes.[4] After that was Islandsdalen, at the end of which was a house where the salted packed cod was exported from. Right in front of that house was the water, where my father used to go to a show in which they would turn over in the kayaks. I was scared when I first saw my father disappear under the water, until I saw that he could turn the right way up again and still be sitting in his kayak. The people would shout and clap. Following the dirt road up from Myggedalen, you came up a hill and passed *Ilinniarfissuaq* (the college), and then passed a house where the flower garden was enclosed by a white fence, because sheep walked around the whole town, grazing. Next to the church was the governor's official residence, and a river ran along the left-hand side of the road. A short wooden bridge led down to the bakery, and next to that was the KGH. The colony harbour is a protected area today, so it looks just like it did when I was a child. After the US Consulate was the municipal building and then the road where the dwarves lived. Agnethe, Karla, and all of their siblings lived further along the road. One got to our house by following the river due south.

On January 7, 1948, Victo and I woke up early and listened to the sounds outside. That night it had been difficult to sleep, because we had never experienced such a blizzard. The red, brown, and green wooden houses now looked completely white, and the snow was falling and swirling so wildly that you could not see your hand in front of your face. It was the Holy Three Kings Day (Epiphany). When we were out playing, we could hardly breathe in the blizzard, and the snow stuck to us, even to our eyebrows and eyelashes. It was fun, but we had to stay right in front of the house, because the snow was falling so heavily, and was quickly forming into large drifts. It was nice and warm in the kitchen and the living room, because my father kept the stoves burning all day when it was cold in the winter. When we reached the kitchen door, Victo and I struggled to get in first. I won, pushed the door open, slumped down into a chair, and clambered up happily onto the kitchen table to open the curtains. Victo came running in after me, and as I opened the curtains, I screamed. In a snowdrift that reached right up to our kitchen window, there was an awful sight – it was a thin lady, wearing a checkered dress, holding a cigarette, and she was completely frozen and had a bird for a head and a black face. I stared at her and kept screaming until my father took me down from the kitchen table and comforted me. It turned out it was a woman who hadn't been able to find her way home in the blizzard – she had fallen down right in front of our kitchen window, and had frozen to death. One of our hanging eiders or grouse must have fallen onto her head during the blizzard. A few days later, she was buried. Mom and Dad obviously knew her, so we attended the funeral, and I saw that some people were crying. I could not understand why anyone would cry over such a monster. It turned out that her face looked black because she had put ashes on it, in celebration of the Holy Three Kings; this was a tradition at the time. My cousins Kristian and Nikolaj were on their way to school that day and had walked past our kitchen door. They had also had a shock when they saw her lying there below our kitchen window.

King Christian X's birthday, which was on September 26, was always celebrated. From 1947 onwards, on March 11, the birthday of King Frederik IX, was also celebrated. The whole town would gather at the colonial port; the governor gave a speech, and there was a good deal of shouting as the cannons were fired, giving off smoke and steam. When they were fired, the cannons would jerk backwards; we would put our hands over our ears, and I always worried that the cannons would roll back into the legs of the people who were firing them. There was great celebration and joy, and after church, all of the married men received a 'King's Parcel', which consisted of a kilo of porridge rice, a small bag of coffee beans, a packet of sugar cubes, a small bag of tea, and a packet of biscuits. Those who had not been to church were given named tokens for their packages, which could be collected later at the KGH. On these afternoons, Mom would bake a Greenlandic cake. It was fun to watch as Mom took her little coffee grinder down, popped the coffee beans from the 'King's Parcel' in, and started to mix the ingredients together. It smelled lovely; and later on, we would celebrate the king's birthday with coffee and cake at the table in the living room. As it was a solemn occasion, a small Danish national flag was placed on the table. Although I could not imagine what a king looked like, I felt that he must be an important person, since everyone celebrated his birthday. For Christmas, the king gave each of the children a small bag, with either an apple or an orange, and raisins and prunes.

When the American warships came to Godthåb, the roads were filled with soldiers. My cousin Ane-Johanne remembers running after the soldiers, shouting, '*Gi' maj tikkum!*' ('Gimme chewing gum!'). For many years, the children thought that they had been speaking English, but it turned out that it was Danish. They got long flat pieces of chewing gum in yellow wrappers.

Holidays and Holy Days

Mom was busy packing our clothes, and I wondered aloud why she doing it. She told us that we were going on a summer holiday, at Grandma's house. We would have to travel by motorboat for many hours, and then we would get to a place called Maniitsoq. My mother and her big sister had been born there, and my grandmother, aunts and many of my mother's cousins and their children still lived there. It was great to sail out; there were many mountains to look at. When we got to Maniitsoq, lots of people came out to welcome us. For the last part of the crossing, we travelled by rowing boat, and then we walked up to my aunt's red house, where we were going to stay for the holidays. It was scary, because there were loose dogs all over the place. I stayed close to my mother.

My grandmother was very old; she was not well, and she stayed at home all of the time. I was happy about that. I liked sitting by the open windows, and looking out over Maniitsoq, with its many small bridges. We didn't have as many bridges at home in Nuuk.[5] I didn't want to play outside, because there were a couple of dogs next door, and I didn't dare walk past them. It was strange to hear my mother being called 'big sister', and my father being called 'Little Hendrik'. We also went out sailing. The adults caught cod, and then they filled pots with seawater, and cooked the cod over a bonfire. We helped to collect heather for the fire, which burned with a thick smoke. When it was time to eat, they called out to us, poured the contents of the pots out over a large flat stone, and we ate the fish with our fingers. It was nice, sunny day when we sailed home to Nuuk again. I enjoyed getting home, because there were not as many people inside, and no dogs outside.

One could buy jam, which came in large tin cans, which looked like gold. When the cans were empty, we brought them out into the mountains, where we used them to collect the blackberries we picked in the month of August. One day, my big sister and I were allowed to go far up into the mountains with some playmates to pick the blackberries. We walked past Dad's workplace, past the football field, and into the mountains. There were plenty of blackberries. It made a hollow sound when the blackberries landed in the empty can. We went further and further into the mountains, and in between them, we could look out at the bay at Kuløen. The sun was shining, and warming us nicely. Suddenly, someone was shouting, 'A polar bear, a polar bear!' The boy was pointing down towards the bay, and there was a polar bear there. The big kids ran home, screaming; I dropped my can of blackberries, and ran as fast as I could, all the while shouting at my big sister, 'Wait for me!' She just carried on running, with me crying after her. It was a long way to run, and when we finally got home, some hunters who had already heard the rumour were rushing up into the mountains, in order to catch the polar bear.

On New Year's Eve, there was a Greenlandic ball game. The ball was home-made, from fabric, and all the men in the town fought, in two opposing teams. One of the teams consisted of those who lived to the north, and the other consisted of those who lived to the south. They started at Islandsdalen and fought their way through Godthåb, out onto Skibshavnvejen. There was a lot of scrambling and shoving, and the ball was ripped, sat, and lay down upon. The men fought on, and there were plenty of us youngsters, as well as the wives, who were excitedly watching the fighting. The men laughed, screamed, and shouted at each other. Eventually, the ball was completely torn apart, and I never found out which team won.

On Sundays, the whole family went to church. I sat admiring the angels and gasped at the sight of Jesus on the cross. Victo and I prodded, tickled, and giggled at each other, until we were told to hush. Sometimes, we went over to visit my Aunt Sofie, her husband, and my cousins after church. Other times, we went over to Grandpa's.

Mom was busy when we were visited by strangers. My father could speak several languages, so when there were Danes, Englishmen, or Frenchmen in town, they were invited to our house, because Dad was one of the few Greenlanders who could speak with them. It sounded weird to us when my father talked to them. We giggled and ran around until we were told to go out and play. I learned my first Danish words from these guests. Mom told me that I used to boss my little brother around in Danish. '*Hans, kom nu, spis. Nej, du ikke løbe*' ['Hans, come on now, eat. No, you mustn't run'], I commanded him.

Another summer, we went on holiday to the fjord at Qooqqut. We travelled by motorboat in the beautiful weather; the sun shone from a cloudless sky, and the sea was like a mirror. In Qooqqut, there was a wide sandy beach where we played. We had to take off all our clothes; the water wasn't that warm, but we had new rubber boots, so we played and ran around, wearing only our rubber boots. We were with our cousin Bibi, and we ran up and down the sandy slope, collecting seaweed and dry reeds. When we needed to pee, we simply sat down in the water. It was fun. My father fished, and in the evening we ate either the cod he had caught in the sea or the salmon he had caught in the big river. When we played in the mountains, the mosquitoes annoyed us – very itchy. Mom managed to fit in some riding, on one of the Icelandic horses in Qooqqut. My dad brought his nice camera. When one looked down into it, it looked as if the people one was photographing were upside down. We could only see his bottom half when he was about to take our photographs. We had to stand completely still until – click! – he had taken the picture (Figure 4).

Back in Nuuk, Victo and I were sitting outside one day in the lovely weather, eating a snack, and we saw smoke coming from one of the houses up on Radiofjeldet. We shouted to our mother, '*Takuuk!*' ('Look!'), and she exclaimed, 'Oh no! It's burning!' Suddenly, flames were bursting out of the house's windows. The residents got out of the house very quickly, and several people rushed to help put out the fire, but the task was impossible; they only had a single barrel of water, next to the stove. The fire made a big impression on me; it was right next-door to our cousin Ane-Johanne and her siblings' house, where we often played. The next big fire we saw was when the transmission station, which was surrounded by a lot of masts, burnt up on Radiofjeldet. The boys from the town, including my brother's playmate, Ejvind, eagerly helped to put out the fire, by running backwards and forwards with tin buckets of water.

FIGURE 4   Holiday in Qooqqut. Mum holding Hans with (left to right) me, Victo, and cousin Bibi. Dad took the photograph

The US Consulate had the first car in Godthåb, and once a week, in 'Amerikanhallen' ['the American Hall'], the adults could watch the 16 mm feature films. It must have been a great experience – just like sitting in a real movie theatre. Karl O. Egede was the director of the cinema in the old barracks.

In 1949, my father bought a broomstick. He hammered crowberry and small birch branches onto it, and Mom decorated it with homemade candy houses, and on put a homemade star on the top. This is what our first Christmas trees looked like when I was little. It was an exciting time. On Christmas Eve and Christmas Day, the big kids walked around the houses singing, carrying homemade lanterns with Christmas tree lights on. By six o' clock in the morning, the young people and the adults were already out singing, and there was a church service at eight o'clock, where we always sang *'Juullimi iivangkiiliu'*, *'Qaammarpoq Nuna'* and *'Illunguaqarpoq tappavani'*. In the days between Christmas and New Year, our friends were invited over to see our Christmas tree. My big sister's classmate, Steffen Heilmann, has told me that this was usually between two o'clock and four o'clock in the afternoon. First, the guests would sing a Christmas song outside the door; then, they would come into the living room, and we'd all dance around the Christmas tree, singing *'Kaavitta, kaavitta, kaavinniarta'* and *'Immakullu nunakkullu'*. Then came the big moment, when we each got a trip into the storehouse – and in there were prunes, raisins, and candy. It was really special if a ship had docked before Christmas, because then there might even be an apple in the storehouse.

My Father Dies

Godthåb Hospital was built in the 1930s, and my father had been hospitalised there. It was quite modern, with electric light, central heating, a good water supply, and a basement running along its entire length. In the autumn of 1950, my father, who was thirty-one years old at the time, started to cough a lot. He had pain in his chest, and he started to get weaker and weaker. By his birthday, which was December 21, he couldn't even manage the stairs up to the bedroom, and so he had to sleep on a mattress on the living room floor. We children could sense the sad mood that was in the house; our mother was very upset, but we didn't really understand why. When the family came to visit, we had to be quiet. We were allowed to sleep in Mom's and Dad's beds, alternately, and one evening I lay thinking about when my father would recover. Every morning I woke up excited, wondering if today he would be feeling better. But he just lay there in bed the whole time.

One night, Mom came and woke us up, and helped us downstairs. She whispered to us that we should come down to the living room and say goodbye to our father. It was January 23, 1951. I wondered why we had to say goodbye to Dad in the middle of the night. When we got down to the living room, he called to Victo, in a low voice. He tried to lift his head from the pillow, but he couldn't. Victo had to lean her head all the way down to him; he whispered something to her and kissed her goodbye. Then it was my turn. He took my hands and whispered, 'Goodbye, my girl. Now, you must promise me to be good.' I nodded; he kissed me, and a solemn mood enveloped me, giving me goosebumps. Then my mother took our four-year-old little brother by the hand, the tears rolling down her cheeks. They squatted down next to the headboard of Dad's bed, but he couldn't even whisper, or make any gesture now. My father's siblings were gathered around the living room and in the hallway, with handkerchiefs over their mouths, and they nodded solemnly to me and Victo, as we were sent out into the kitchen. We stood there, completely lost, and looked at each other with tears in our eyes. Victo started sniffling and was completely silent for a long time. Then we heard the front door open, some men's voices, and then a lot of steps. Soon after, they walked past the kitchen window, with a stretcher covered in a green blanket. Our father was lying on top, completely still, and four men carried the stretcher. Victo and I looked at each other, then we went out into the living room to Mom, who was standing there with Hans in her arms. Crying, we embraced her. Everyone around us cried just as quietly.

Later on, Ane-Johanne told me that Victo was late for school on the morning that our father died. When the entire second grade was seated, Mr Forchhammer, the Danish teacher, asked them, '*Hvor er Victoria?*' ('Where is Victoria?'). At first, no-one could explain it to him in Danish, but then the most courageous of the students, Kaaleeraq Heilmann, stood up, walked to the front of the class, and said in Danish, '*Victoria, far*' ('Victoria, father'), then lay down on the floor, pretending to be dead, whilst saying in Greenlandic, '*Unnuaq toquvoq*' ('He died last night'). There was complete silence in the class. Kaaleeraq got up, and went back to his seat. The class teacher, who understood the seriousness of the situation, took a piece of chalk, and slowly and solemnly wrote on the board: '*Victorias far er død*' ('Victoria's father is dead'). Then he turned to the students and said, slowly, '*Victorias far er død*'. Then he handed out a piece of paper to each student, so that they could write down the same sentence that was on the board: '*Victorias far er død*'. Ane-Johanne has never forgotten that; it was the first sentence that she learned in Danish.

A few days later, Mom told us that our father was going to be buried. She was busy finding our nicest clothes and our handmade coats. 'We have to say goodbye to Dad for the last time', Mom explained. I didn't understand any of this; and when we finally reached the church, we walked straight past it and went further on up to a little red house. It was the chapel of rest. I was curious about what would be in there, which quickly became clear. We were the first to be lifted up; my father was lying there, pale and completely still, as if he was sleeping in the coffin. Mom whispered to me that I should kiss him for the last time. I pursed my lips and gave him a kiss on the forehead. As I kissed him, I stiffened in shock. 'He's ice cold', I thought. I was then sent outside the chapel of rest to wait, until the whole family had kissed him. Then the casket was sealed. It was then that it dawned on me that I would never see my father again.

It was strange to sit in the church and to look at the white coffin with its lid on. 'Our Dad is inside – they've locked him in', I thought. The people were sniffling and coughing in between the hymns. Then some men came in, and carried Dad's coffin out of the church, putting it on a trolley, which was covered with a white sheet. The men slowly pulled the trolley with Dad's coffin down the last long road along Skibshavnsvej. Many people walked behind the coffin. We walked past KGH, past our house, past the dentist's house, down the hill, and then I could see what they called '*den grønlandske kirkegård*' ('the Greenlanders' cemetery'). In the middle of the cemetery, there was a large hole in the snow. My father's coffin was put into the hole, the priest threw some soil on top, hymns were sung, and we said *Ataatarput* (the Lord's Prayer). I folded my hands very quietly, and I felt completely abandoned in the cemetery, thinking, 'Will I never get to see my Dad again?' Afterwards, the whole family went home with us and had coffee, tea, and Greenlandic cake. Aunt Sofie had baked a big cake, too.

Notes

1 A prisoner population has existed in Greenland from the beginning of Danish colonisation (1721). Prior to that, there had been no (Inuit) history of physically incarcerating wrongdoers in Greenland. However, after the Herstedvester prison in Denmark opened in 1935, Danish authorities sent those they deemed as serious offenders in Greenland there.

2 The unit of Danish currency is the *krone* (plural: *kroner*), which divides into smaller units of (one hundred) *øre*. In March 2022, the exchange rates were 6.73 Danish

kroner (DKK) to the US dollar, DKK 5.33 to the Canadian dollar, DKK 7.44 to the Euro and DKK 8.86 to the British pound.

3 Founded in 1774, *Kongelige Grønlandske Handel*, or KGH (the 'Royal Greenland Trading Department'), was a Danish state enterprise which as well as exerting an absolute trade monopoly, managed the government of Greenland until 1908. The monopoly lasted until 1950, when preparations began for the political integration of Greenland into the Kingdom of Denmark. Following the introduction of Home Rule in Greenland in 1979, KGH split into a number of successor companies.

4 The Danish words for 'mosquito' and 'valley' are '*myg*' and '*dal*', respectively. Hence, a literal translation of this place name is '(the) Mosquito Valley'.

5 Since its foundation in 1728, Nuuk has been Greenland's largest settlement. Prior to Greenlandic Home Rule in 1979, when Greenlandic place names became official, it was usually known by its Danish name, Godthåb. Helene uses both versions of the town's name in her text, and I have followed her uses in this translation.

3

DENMARK, 1951

In Care

It was then deemed time for us to go into care in Denmark. Save the Children had found foster parents for us, and as my eczema had not improved, I was to be placed with Dr Jørgen Kringelbach, in the Copenhagen suburbs. I was picked up by a lady from Save the Children, and we travelled by train towards Copenhagen. As we approached the city, I could see a lot of houses; my hand was being held, which was probably just as well, because I was gasping at the sight of the tall buildings. A car came to collect us at the train station, and I was driven out to a red, single-storey brick house. Out there, the houses were not that tall, but there were lots of cars on the road. The father in the house wore glasses, the mother stayed in the bedroom, and the daughter was the same age as me. The father examined my elbows and the backs of my knees, after which he rubbed them with thick, black thick ointment, which looked like tar. Then he explained to me that I wasn't allowed to sit on the armchairs or the couch in the living room.

I was unhappy. 'Why do I have to live here now? When can I go home?', I thought. I didn't like being there. I wouldn't talk to my foster parents – I only nodded or shook my head when they addressed me. I only talked a little with my foster sister. At meals, I had to finish everything that was on my plate. It was bad when we had spiced sausage. I looked down at the food for a long time and I took a bite that seemed to swell in my mouth. 'Eat up, now', said the foster mother, who explained to me that it was her mother who had made the sausage. That didn't help. I had to sit there for hours. One day, my Aunt Kristine, and my cousin, Theodora, visited. I was so happy that I cried for joy. I told them about the sausage, but they couldn't do anything about it.

When I went out to play whilst I was with the Kringelbach family, they said to me, 'Don't go out on the road'. 'No', I said, and they repeated, 'Don't go out

DOI: 10.4324/9781003241843-5

on the road', and I repeated, 'No, no'. I should probably not go out on the stupid road. Then I said to myself that I'd ended up fastened down like a nail, behind their garden wall, and I must not do anything wrong. Their neighbours were the Feldbo family, and one day, when it had been snowing, the father came over with a toboggan, and we went sledding in their garden. Those were lovely moments, and he explained to me that one of the birds was called a 'robin'. I thought that that was a weird name. Every day, the wife lay in bed, and one of the children said that she had eaten nails. I thought that was probably why she was so weird. I only ventured into the bedroom when I walked past. Inside the daughter's room, we learned how to knit, but I found it awkward.

It was Christmas Eve, and the Christmas tree had real branches, and it smelled lovely. It was much nicer than the one my Dad had made. My foster sister got a very big doll for Christmas, which was really nice. I got a little doll. I was angry at my foster parents, because they forced me to eat, so every time they tried to ask me something, I pretended that they weren't there. In the early spring, when my eczema finally healed, I moved again. When I was collected, I thought that I was being taken out to a ship and that I would soon be home, but we drove further and further out in the countryside, down to Brøderup primary school at Præstø Fjord. It was a long drive. It looked very big and nice, and I was totally amazed when we drove along the entrance road. To the right, there was a small pond, and next to it was a semicircular hedge, which surrounded a bench, and a round granite table. From there you could enjoy the view of the pond with goldfish and waterlilies. In the middle of the pond, there was a naked boy peeing, and I was a little embarrassed when I first saw him. There was a huge lawn, with a long flag-pole on one side. Opposite, there were two short flagpoles. Behind the lawn was a large area with various vegetables and plants, and out towards the main road, there were fruit trees. We parked in the courtyard in front of the dining room.

At Brøderup primary school, the principal couple were Ingrid and Ejner Greve. Their own children were Vagn, aged fourteen; Gunvor, aged twelve; and Tove, aged eleven. The couple had taken care of children for many years – from Austria, Hungary, and Schleswig. One day, they had been contacted by the Save the Children, who asked them if they could take care of a 'problem-child' from Greenland. They had been told that, 'She is seven years old, and her name is Helene. She will be here for less than a year'. They had taken care of prob-lem-children before, so they said that they would give it a try. My new foster family welcomed me; they were smiling and looked nice when I arrived. Behind them was a winding staircase up to the first floor, with paintings on the walls. I was sullen and self-contained when I arrived, but with so many positive people – students, teachers, kitchen girls, and my foster family – around me, I quickly thawed.

'The problems with the "problem-child" were overlooked', was how Vagn expressed it to me many years later, at a Christmas party at the family's home. It was a lovely place that I had come to, full of joy, and with plenty of space. I was given a room behind my foster parents' bedroom, which opened out to a loft. On the opposite side of my room, the maids and the kitchen girls had their rooms. They were very kind, and I often visited them, or went through their rooms, to go downstairs, and out into the yard, or into the kitchen to get a treat or something to drink. The family had a small white poodle named Vips, which was friendly. In the beginning, I was scared of the dog, because when I ran, it would run after me and would nip me on the backside. One of the first times that I was in the car with my foster mother, Vips jumped in. I was so startled that I jumped over to one side of the car, and the dog jumped after me. I moved over to the other side of the car, and again, the dog jumped after me. This continued until Ingrid, laughing, picked up the dog and put it in the front of the car. Then I could breathe out, with relief!

I quickly became friendly with all of the young people at Brøderup primary school. I was with them a lot and became everybody's little pet. It was great fun, being thrown backwards and forwards by the students, when they were in the mood. It was a little difficult not to disturb them for hours. Sometimes they opened the window, called me over, and asked me, in a whisper, to fetch them an apple. It was great when there were 'housewife days' or 'student days' at the school, with gymnastics displays, and music on the big lawn. The students danced for the guests in the evening. After the dance in the hall, there were fireworks, and there was singing around the illuminated pond, which was filled with beautiful waterlilies. Imagine – flowers in the water! I never ceased marvelling at this. Every time guests came to the school, they signed a guestbook. I also wrote my name, 'HELENE KRISTOFFERSEN', but I wrote the S's the wrong way round. Underneath my signature, an adult added, 'Godthåb, Greenland'.

At the gable at one end of the large student building, there was a very tall tree. Up in the tree, there was a firmly tied rope, so we had a swing, on which we go really high. Inside the building was a nice, big living room. When you went from the hall into the living room, there was a large desk, and to the right of that, a two-seater sofa. Next to that was a huge corner bench, upholstered in striped fabric, and then a table with winding legs. In front of the patio door were a couple of armchairs and three small pull-out tables. Behind an opening in the wall – which was the width of a double-door – lay the fine parlour, with its beautifully upholstered sofa in dark wood, with an oval, lustrous, and ornately designed coffee table, which had patterned edges and curved legs. There were also large paintings, an ashtray, a standard lamp, and a large grand piano in one of the corners. In the back corner, there was a door to the kitchen, and opposite that, a door to the room with a fireplace. The student rooms, which were in the large, white student building, were accessed via the room with the fireplace.

Ejner and Ingrid had an exciting surprise for me one day. It was a light blue girl's bike. My foster siblings had bikes, so I wanted one – and I got one. 'How on earth do you balance on this?' I wondered. I was told that in order to ride it, I'd have to learn how to balance – and whilst I was learning, every time I didn't fall off I would get ten øre. I practised, fell off, practiced some more, and fell off some more, but at last, I got my 10 øre. When I could ride safely, I would go back and forth all day, alone on my bike.

Brøderup primary school was located off the busy Highway 2. The blacksmith lived on the same side of the main road, and the Karlshøj grocery was a little further along. So I was allowed to stroll alone along that side of the road as far as the grocery, but I had to promise never to cross the big road. The blacksmith had a dog that was black with brown spots and had a very small head. Back then, Dobermans were generally referred to as 'police dogs', because the police used them. The couple who owned the grocery, Mr and Mrs Hansen, also lived on the road. I don't remember now if it was them who had a German Shepherd dog, but I do remember that a big dog chased me one day when I was racing into Brugsen in order to spend my 10 øre. I got a fright, and I ran across the main road. Later on, I had to go back across the main road in order to return to the Brøderup primary school. I did this without really thinking about it, so the next day, I decided that I would act more deliberately. I didn't say anything to anyone, but I looked really hard before I crossed over to the other side of the road. When I got as far as the dairy, I went back across the main road to the Karlshøj grocery and started trading with Mr Hansen. 'Hello. What can I get you?' he asked. I wanted a cream bun. I showed him my 10 øre coin. 'Yes, would you like anything else?' asked Mr Hansen. I looked around the shop and I noticed a skipping rope. I pointed it out, and he put it on the counter. 'Yes, would you like anything else?' he asked again. This time, I noticed a bucket, so I got that, too. On the way home, I thought about how much you could get for 10 øre. You couldn't really do that, of course; but Mr Hansen couldn't resist the little, dark-haired Greenlandic girl. My foster parents got along very well with the couple from the grocery, and the manager of the dairy and his wife, too.

Ejner used crutches. I came up with different ways to grab at them, so that he'd tumble over. My foster siblings would sometimes say something nasty about this, but my foster father's indulgent remark was, 'It does not matter. You always have to support those who are struggling.' That was his motto.

The school was a community in itself. Whilst I was everyone's 'pet', I ran around the student wing and also in the teachers' areas. If I didn't get my own way in one place, I made sure that I got it somewhere else. They treated me like that because they thought everything was sweet in my life. Tove told us that, later on. My foster siblings were brought up quite freely, but they were never in any doubt about what they were and weren't allowed to do. They knew the limits. I also learned about the boundaries, but did not always adhere to them. To give some examples, Ingrid and Ejner's pre-dinner nap could not be disturbed; and the house could be burning down before my foster siblings would dare to wake their parents up. 'Well – I dare', I said – and I did. Nothing happened to me, except it being taken as me having fun. I am surprised that my foster siblings never got jealous of me.

Gunvor was often ill, and when he was, he had to stay in bed. One day when I had been up entertaining her, I went downstairs and said to Ingrid, 'Gunvor is calling for you'. My foster mother rushed up the stairs and asked her what the matter was. 'Why are you asking me that? I wasn't calling for you', replied Gunvor. I had felt so sorry for Gunvor, that I wanted to get his mother to look in on him. 'What is Vagn doing?' I thought, the first time that I looked for him. I found that he spent his time doing his homework in his room, with the big black-spotted dog, Max, for company. The one who spent the most time playing with me was Tove. We would play with our dolls, either inside or somewhere in the huge garden. One day, the family was visiting my foster mother Ingrid's family in Jutland. They were leaving for three or four days, and there was no room in the car for anyone apart from Ejner, Ingrid, Vagn, Gunvor, and Tove. So I had to stay at home, and be looked after by the Simonsens, who were a teaching couple, and lived in one of the school's teacher residences. I was deeply offended by the unfair plan. Through the car window, I stared at Vagn, and as I stood waving at them, I shouted, 'Why does Vagn get to go, and not me? He'll only sit there reading, anyway!'

One day, a girl my age came over to play. We were rushing around having fun, but then she needed to pee. She asked me to go with her to the toilet in the garage. I didn't want to, but I went along anyway. When she had finished peeing, she told me to pull my pants down. I didn't want to do that. She said, 'Yes, but we have to see if our *tissekoner*[1] look the same. To keep her quiet, I quickly pulled down my pants, and quickly pulled them up again, just as soon as she had looked. Then we ran outside and started playing again.

I was spoiled by my foster parents. One day, when my foster mother was going into town, she asked me if I wanted her to bring something back for me. 'Yes, a china doll, like Tove's', I said, without looking up. A little while later, she came back from town, with a large china doll with brown, curly hair, and brown glass eyes that opened and closed. The doll was dressed in a flying suit. It was cute, and I immediately named it after my youngest foster sister, Tove. I was allowed to borrow Tove's toy pram, because she was at school a lot of the time anyway. As time went on, the long trousers that I had brought from Greenland became quite worn out. My foster mother was usually very particular about our being dressed properly, but she didn't say anything about my trousers. Later on, Tove told me that, in her opinion, my foster mother felt that I was probably emotion-ally attached to the trousers – a reminder of my mother. So I was always allowed to run around in them.

My foster siblings had been told that I was going back to Greenland to live in an orphanage. They had objected to this, reasoning that if I was going to live in an orphanage, their family might as well keep me. My foster parents told them that I had a mother at home, and that 'you can't just keep other people's children like that'. My foster siblings understood that, of course. They thought that I needed to go into an orphanage because my mother couldn't take care of me. They knew that in Denmark, children went into the care system, for either short-term or longer-term periods, when their parents could not care for them. After the war, my foster parents fostered several children through Save the

FIGURE 5   Eva in Denmark, 1952

FIGURE 6   Gâba (front) in Denmark, 1952

Children. These were children who had been injured or orphaned during the war years, and my foster siblings imagined that something similar had to be the case with me. They had heard that there was tuberculosis in Greenland and that many of the children there were malnourished.

The Return Journey

On September 25, 1952, I was collected by a lady from Save the Children. It was sad to say goodbye to my Danish family. I didn't know if I would ever see them again. I had a lump in my throat; but on the other hand, it was incredible that I was finally able to go home to my own mother, and see my big sister and my little brother – home to my beloved Greenland. I had very mixed feelings. The car was loaded with my bike, my lovely china doll, and everything else that I had acquired during my time at Brøderup primary school. Tove had given gave me the doll's pram the day before I went home, so I brought that with me, too.

There were many people at the quay in Copenhagen harbour, where the *MS Umanak* was ready to take us home. It was very confusing, because in the crowd, there were some photographers walking around and taking pictures. In the text underneath a group photograph, it says (Figure 7):

> There was a big goodbye at "Umanak"'s departure yesterday. The sixteen Greenlandic children, who have been in Danish care for a year-and-a-half, returned home to Greenland. Pictured (from the left) are the business manager Sckerl, with his daughter Mette, and his son Ole, who will say goodbye to their foster-siblings, Karl and Anesofie. Karl is too young to understand a great deal about it; but for the bright eight-year-old Anesofie, it was sad – she kept hiding her face behind the railings, and wiped her eyes whilst her friends on the quay tried to comfort her. Next to them is the manufacturer, Hans Sachs, with his baby boy, Aron, and the hairdresser, Mrs. Ketty Rasmussen, with Gabriel (who is known as Gabbe), who had such difficulty leaving that Mrs. Rasmussen accompanied him as far as Helsingør.

The lady from Save the Children took me to meet an old lady in a green uniform, who had a big smile on her face. 'This is your new mother; her name is Miss Bengtzen', said the lady from the Save the Children. I stopped, looked the old lady up and down, and thought to myself, 'What? Our new mother? That can't be right!' I must have looked extremely surprised because as far as I knew, I was on my way home to my own mother. In the midst of all this confusion, I saw Agnethe, and the twins, Eva and Marie. We ran over to the big ship, and up its gangway, and stood by the railing looking out at Denmark for the last time. In my mind, I sang, 'Now it's time for me to go home to my mother. Mom! Mom! Mom!'

FIGURE 7 Setting off back to Greenland on MS Umanak, 1952

Finally, we sailed away. There were some smaller boats sailing next to ours; Gâba was in one of these smaller boats. His beautiful, blonde foster mother accompanied him as far as the last port, Helsingør, before joining us. The *MS Umanak* whistled several times, and we waved, in the way that we had learned to do so. When we could no longer see land, we were shown down to the cabins, with our suitcases. I had my china doll under one arm. This time, sailing was fun, because we knew that we were sailing home. We were shown around the big ship. Inside, it was very similar to the ship that we had sailed down to Denmark on. On one of the first days at sea, we all had to meet up on the deck for a rescue drill after breakfast, so that in the event of the ship capsizing, we would know what to do. We all got to put life jackets on – it was pretty exciting. We all looked at each other, standing there in our life jackets; but when we tried to run around whilst wearing them, we couldn't manage it. But we could go in and out of our cabins as we pleased. I often went down to check if my china doll, Tove, was still in my bed.

One time, I heard a sudden, loud cry of pain when a door slammed. We ran out from our cabins in order to see what had happened. It was Little Kristine, who had caught her finger in a door. It was so badly crushed that her middle finger was hanging off her hand, like a piece of raw meat. On hearing Little Kristine's cries of pain, Miss Bengtzen (our 'new mother') and the other adults hurried to see what had happened. Miss Bengtzen was an experienced senior nurse, who knew lots of quick fixes. As soon as she had examined the injury, she quickly took Little Kristine into her cabin. Miss Bengtzen found two small ice-lolly sticks; she placed one on top of Little Kristine's injured finger, the other underneath, and then wrapped some gauze bandage around the finger and the sticks. This whole episode happened because Little Kristine, who shared a cabin with the twins, Eva and Marie, had seen them unpacking some of their gum. Little Kristine had followed them and asked them if she could have a piece 'No!', the twins had harshly replied, every time Kristine asked. Kristine had become really annoyed, and had stormed out of the cabin, slamming the door behind her. The doors on the ship were heavy, with sharp edges, and Little Kristine's poor middle finger ended up in a real mess.

Note

1 This word, which is used in the plural here (sing., '*tissekone*'), is one which Danish children (and the adults around them) use to refer to female genitalia. It derives from the Danish words '*tisse*', an informal verb meaning 'to pee', and '*kone*', which means 'wife' or 'lady'. As the literal translation 'pee-lady' is incomprehensible, and there is no direct equivalent that might be understood across English-speakers, I have left this word untranslated.

4

GODTHÅB, 1952

Home Again

It is October 6, 1952. I wake up, feeling anxious all over. Today is the day when we go back to my hometown of Godthåb. As soon as I have washed and brushed my teeth, I pack my suitcase, so that I can be ready to leave the moment that the ship docks. I walk over to the ship's dining room with my coat over my arm, but there is only one waiter there, walking around in short white jacket, and dainty black and white trousers, and starting to set the tables. As breakfast is not ready yet, I head out on deck and see if I can recognise the mountains. Agnethe is constantly running after me; we smile at each other, and gaze out over the horizon. 'Look, there's land!' we say, in chorus. It's nice that it's so cold. We run to the breakfast table as soon as the food is served. I can hardly think of anything other than the fact that today is the day when I will finally get to see my mother, my big sister, and my little brother again. 'I wonder if I will know them?', I think. My whole body is trembling with anticipation, and inside my head, I hear myself repeat over and over, 'Mom, Mom, today, today, I'm on my way home, back to my beloved hometown, Godthåb'. After breakfast, I hurry down to the cabin and pick up my small suitcase. The place is teeming with children; we're all busy getting ready. We can see the harbour. 'There's the red warehouse!' When I leave the cabin, I have a feeling of being late for something, because I can already recognise the mountains, Store Malene and Lille Malene. There is already some snow on their tops. I quickly run around to the other side of the ship and see that all of the people who are coming to meet us are standing on the quayside. We wave, with both hands in the air, and lots of the people on the quay wave back. It's a big moment. I get a lump in my throat as we scurry around; we are rescued by the ship boys, who have to open the link to the gangway. It's high tide, and as

DOI: 10.4324/9781003241843-6

the gangway is set into its position, it looks steep. I can hear the propeller turning, and the noise of the ship's anchor, as we dock.

I scan the crowd, frantically – where is my mother? As I move to a new standing place by the ship's rail, we get close enough to the dock for me to be able to see my mother, and Victoria and Hans. Is it really true? Yes – they're standing there, pointing and waving to me. We make eye contact – and my heart swells with immense relief and happiness. 'Mom, Mom, Mom. Are you real? Home to Mom, Home to Mom!', it sings inside me. As soon as it's my turn, I wobble down the rocking gangway. I walk as fast as I can, all the while clinging onto the rope-railing, finally reaching the end and running the last few yards into my mother's arms. I cling to her; I'm bursting with joy. 'Mom! Mom!', I cry, and I begin to tell how much I have missed her and everything that I have seen and experienced in the year-and-a half since we've been apart. Joyfully, I look up at her; and I'm wondering why she hasn't answered me yet. Then, what she does say is incomprehensible to me. I let go of her, look up, and ask her, 'What are you saying?' In a split second, I'm struck by disbelief; I stare at her, and then at my siblings. They're looking at me as if I'm from another planet. I realise that we can't understand each other at all – we're speaking different languages. I start to feel numb, and completely empty inside. 'What's happening?' I think.

As I'm standing there, trying to get over the shock, I feel a tap on my shoulder. 'Come on, Helene. You can say goodbye to your mother, and pick up your suitcase. You need to get on the bus, over there.' 'Why do I have to do that?' I ask. 'Well, now you have to go to the new orphanage, to live with all your friends, and your new mother, Miss Dorothea Bengtzen'. 'What?' I think, the disbelief wounding me in my chest. I press myself closer against my mother's side, but she doesn't do anything to keep me. I'm close to fainting. 'This can't be true! They're taking me away from my mother again! They can't!' I'm in shock; but with steps that feel as heavy as lead, I walk over to the bus. I look for my mother; she's holding my siblings' hands, and she looks sorry. I cry inside, thinking, 'Why is there no room for me at home with my mother? Why?'

I get helped up onto the bus. Miss Ingrid takes my suitcase. I'm one of six children from Godthåb, and we're all looking completely lost. Our mothers go home, with bowed heads, and some of the children on the bus start sniffling. The whole thing is completely incomprehensible to us. 'Where is the orphanage? And what is an orphanage?' I think. My curiosity helped lift my mood a little. We recognise the mountains, and the Marine Station, and its small bay. We travel up the hill, through a rocky gate that has disintegrated, along Skibshavnsvej, and turn towards Østerbro, where the Danish emigrants live in fine, different coloured wooden houses, which have conservatories. We drive up towards the football field, and then we can go no further. Miss Bengtzen and says: 'Look, to your left is your new home, the Danish Red Cross orphanage.' I look at her ugly green woolly hat, and think, 'Well, that's why she's got a red cross sewn on it, as well as on the shoulder of her DRK[1] jacket.' I sit back, and deep inside I wonder, 'Why do I need a "new mother" when my own mother and siblings live in town?' Then the bus stops. We can't drive all the way up to the orphanage, because the last stretch of the road hasn't been finished yet. There is some type of thick covering over the broken pieces of rock on the last stretch of road. We have to carry our suitcases ourselves. 'Look at the flagpole, they've raised a flag!', someone yells. I look up, and see a Red Cross flag flying. 'Oh, they've raised that sort of a flag, instead of a Dannebrog', I think.[2]

In the *Danish Red Cross Journal*, no. 7, 1952, there was a picture of the Chief Officer, G.N. Bugge, with two children – Barselaj, who was dressed in a white anorak, and Ane Sofie, who was dressed her Greenlandic national costume. Underneath the picture, there was the following text:

> Home from Greenland: Chief Officer G.N. Bugge, Chair of the Danish Red Cross's Greenland Committee, has been on a trip to Greenland to see the new orphanage in Godthåb, built by the Danish Red Cross, and to visit a number of places in order to speak with the government doctors and others who are part of the relief work undertaken by the Red Cross in Greenland. The next issue of this journal features an article by Bugge about his experiences in Greenland, and the importance of the Danish Red Cross relief work to the Greenlandic population. Along with this short report, we bring you a picture of Chief Officer Bugge, with two of the little Greenlandic children that Save the Children has hosted here in Denmark, who are now going to live at the Danish Red Cross Children's Home in Godthåb.

The Orphanage

On September 5, the Danish Red Cross orphanage was inaugurated in Godthåb. The Asmussen and Weber company had invited their construction workers and foremen, as well as Councillor Augo Lynge, Municipal Council Chairman Nikolaj Rosing, the police chief, the regional manager, the local doctor, the regional building manager and several others, for a celebratory glass of wine and piece of cake. Asmussen and Weber's manager, Mr Jakobsen, welcomed the guests, stating that they were prepared to speed up the work, as much as was possible. After that, the local doctor, Mogens Fog-Poulsen, spoke, mentioning the importance to the Greenlandic community that the orphanage would have, and thanking the Red Cross for their interest in the completion of the building. Then the doctor, who had Chaired the Danish Red Cross department in Greenland, introduced the newly appointed principal, Miss Bengtzen, and welcomed her into the work of the orphanage. After this, Nikolaj Rosing, who chaired the Municipal Council, spoke, making the following statement:

> I want to say a few words on behalf of the town, and my countrymen. The erection of this building is one of the clearest examples of Danish efforts, in

the already extensive work with the Greenlandic children. Those Danish efforts for the Greenlandic children will be further enhanced by the opening of this orphanage. This magnificent work will gradually open the eyes of the Greenlanders to how much it means for a population to have healthy, well-groomed children, not least in the years when construction work up here, in every field, is in progress. It is good to have new painted houses, good roads, fast motor boats and all of the other nice things, but the most important thing is that the younger generation grows up well-groomed and well-adapted, in healthy surroundings, so that they can more easily enter the new Greenland, that we all are all hoping for. Let this promising building, for many years to come, be an example for many other Greenlandic homes. Let those children who have enjoyed this good care be amongst those who work for the continuation of the child care work shown by the Danes. With that I, on behalf of my countrymen, say thank you very much for this great gift which we now receive from the Danish Red Cross.

The orphanage is a brand new, very large, red wooden house. When we have changed our clothes, we are allowed to look, and run, around the house. First of all there is the entrance, and behind the door to the left, there is a cloakroom with twenty spaces on a bench, which we can sit on when we're hanging up our coats. From the window, we can see out onto the mountains behind the orphanage. The door on the right goes into the headmistress's cloakroom and toilet, and then there's the door that leads into her combined office and bedroom – her desk is at the window, opposite her bed. The next door to that leads into our "new mother"'s living room, in which the large window opens out onto a view of the nearby mountain, and a tall flagpole. There is a door leading out to the large general corridor, which is filled from floor to ceiling with white painted cabinets. In the first children's room, there are four low beds, one pair on each side of the room. At the foot of each bed is a white painted cupboard. Altogether, there are three of these completely identical rooms, and between the rooms, there are several white floor-to-ceiling cabinets. The children's bathroom is on the same floor, which has frosted windows, and small washbasins and mirrors, all along one wall. There is a shower for the adults, and on the other side of the shower is our big bathtub. There's a wooden bench at the end of the room. A door leads out to the two high-seated, and smelly, children's toilets. They are separated by a white wall, which is raised 50 cm above the floor. The toilets have doors with swivel lock. When they are 'occupied', the locks show red; when 'available', they show green. The toilets themselves have black exteriors, with black toilet seats, and black lids, and inside them, there are bucket with handles. When they are empty, a strong liquid is poured into the buckets; it has to get rid of the worst of smells. On the walls, there are toilet roll holders, and behind the doors, there are small washbasins.

The next door from the general corridor leads into the sewing room. There is a long, fixed table, a north-facing window, and a small table at the end of the room, with a new sewing machine. The door after leads into the large, bright dining room. Behind that are sixteen small, white, built-in cabinets, with round, black handles. There's another door, leading out to the utility room, and to the left of it are four more small, white small cabinets. The radio speaker is on top of one of them, and a narrow counter top runs alongside, on the far left of which is a grille, so that the heat can rise from the small radiator underneath. To the left of the small cupboards is a sliding door, designed for serving from the kitchen. There is also a set of French doors, from which you can look out onto the large terrace, and in the living room, there is a piano. The windows are tall, and south-facing; and beneath them are benches, with at least ten compartments, where the communal toys are kept. The other furniture comprises light, wooden tables, which can be assembled as required, and about twenty-five chairs with curved bars in the backrests, and hollowed-out seats. The lamps hang from a ceiling fixture, at the end of which – as well as on one of the walls – the Danish Red Cross flag is hanging. On another wall, there's a framed photograph of the Danish royal couple, with their three little princesses.

On the way out to the kitchens, there is a utility room, which has steel tables and sinks along three of the walls. The next door leads out to the kitchen itself, where a long steel table with two sinks, and wide shelves underneath for the pots and pans, runs the entire length of the north-facing window. To the left of the door to the hallway is a huge black wood stove, with several hob rings, and a large oven. Next to that is an electric stove, and the biggest mixer that I've ever seen is attached to the floor nearby. There are cupboards and drawers with dishes, bowls and cutlery, above and below the serving hatch. We run onwards to see the next room, which is the larder. There are floor-to-ceiling shelves filled with flour, sugar, salt, and tinned fruit. Then we go downstairs to another food storage room, which has a low ceiling, and is lined with wooden shelves, with butter tartlets, potatoes, apples, carrots, and preserves. It smells wonderfully of apples. Next door to that room is the basement, with its coal cellar and boiler room. There's a big industrial washer and clotheslines everywhere; the basement windows were low, and its floors were made of cement.

We walk up all of the stairways, until we reach the large top floor, with its sloping windows. Here, there are two smaller children's rooms, and across the hallway, a big room, for four children. There are built-in cabinets on the wall on each side of the door, and there is a table in the middle of the floor, with four chairs. From this floor, one has the best views onto the marsh below, and of Store Malene; and to the right, one can look out over the town of Godthåb. I look out in the direction of the mountain, and I can see where my family home is. I feel upset about this, and I'm just about to burst into tears, but the others call, 'Come on! Let's see the rest of the orphanage'. The next rooms on the attic floor are for the child nurse, and the two *kiffaks*.[3] The last room is similar to the large room for four children, and it is located directly above the headmistress's apartment. The children's toilet is next to Ole's and Eli's room, and is just like the one downstairs, and there's a children's bathroom with four washbasins. The mirrors are positioned high up on the walls. Next to that is the adults' bathroom, which has two washbasins, and a door leading to the adult toilet. On the stairway, there is a door with small windows, and next door to that was the best room in the whole house. It's a storage room, lined entirely with shelves, and going in there is like going into a toy store. It is packed with gifts, for upcoming birthdays and Christmases. A small step ladder leads up to the large, cold attic. Next to the gift room was the last top floor room, which had a frosted glass window in the door. This was the medical room, which had a hospital bed, and a glass shelving unit, filled with gauze, patches, and syringes. There was a weighing scale on the floor, a measuring rod fixed to the wall inside the doorway, a small desk with a chair in front of the window, and an armchair in one corner.

Benze

Downstairs, a ship's bell is being rung, and we are rush down to the dining room. We are told to find a seat and to sit down. We're noisy, but when Miss Bengtzen quietly touches the ship's bell, we fall completely silent. 'Thank you', she says, 'You should be quiet when I want to say something to you'. We are told that the rooms are going to be allocated. Eli, Ole, Albert, and Aron will be in the large room above the dining room; Eva and Marie, Big Kristine and Bodil will be in the other large room; Agnethe, Ane Sofie, Little Kristine, and I will be in the first room on the ground floor; Barselaj, Gabriel, Søren, and Little Karl will be in the room next to that; and the last room will be reserved for the new children. I am thinking to myself, '"New children"? And where are the other children, who went to Denmark with me?' when Little Karl starts sniffling, quietly calling, '*Aaqa*' (big sister). A decision has been made that he will not be in the same room as his older sister, Ane Sofie. Everyone looks over at the two siblings. We are all feeling tired and upset with the whole thing. It is getting more and more difficult to control the lump in my throat, which is getting bigger and bigger. Miss Ingrid rushes over to Little Karl and puts him up on her lap. 'When the bell rings, we'll eat. Please bring your suitcases up to your new rooms, and unpack'. I choose the first bed on the right, and the first cupboard on the right, and unpack my clothes and toys. We ask where we should put our empty suitcases and are told, 'Just put them under your beds for now'. As soon as Ane Sofie has finished her unpacking, she goes along to help Little Karl unpack. Agnethe and I walk around and look at where the other children's rooms are; we are most interested in the other children from Godthåb. We hold onto each other, and asking in whispers about why we aren't allowed to go home to our mothers.

There is a sharp clanging of the bell, and we run into the dining room. Miss Bengtzen asks for quiet whilst she tells us the names of the different adults. Miss Blom is a Danish child nurse; Miss Ingrid Holm is a beautiful, Greenlandic teacher, who is very good at Danish; Miss Karen Ludvigsen is a Greenlandic cook, who is not very good at Danish; Mrs Gertrud Boasen is a seamstress; Miss Birgitte is a *kiffak*; and there is a male janitor named Jens, who has to empty the latrines. Some of us hold our noses, but we get a sharp look from Miss Dorothea Bengtzen. We find her name difficult to pronounce, and quickly, she lets us call her 'Benze' instead. Finally, we get to eat, but we only get porridge, and the milk is in steel jugs. In Denmark, we had got used to milk bottles. We are told that no shopping has been done yet and that the only thing in the house is oatmeal. When the KGH store opens tomorrow, some other food will be bought.

We are then introduced to the table etiquette, which we have to follow. First, we must wash our hands before every meal. Then there was a break. One of the children was sitting there, picking his nose. 'Once you've washed your hands, don't pick anything!' We giggle, and we can sit wherever we want. 'Straighten your back, stand still, look straight, put your hands up on the table edge, and wait for someone to give you permission before you start eating. When we finish eating, we must sing, 'Thanks for the meal, thanks and you're welcome, it was good, the plates are empty, thank you for the meal'. Miss Ingrid plays the piano, and we sing it through several times, until we have learned it. Benze also tells us, 'Before going to bed, you have to pee, brush your teeth, and fold your clothes neatly, putting them on the stool next to your bed. At 8 pm, Miss Blom will arrive. You then have to kneel down on your beds, and sing, 'I'm tired, and

FIGURE 8 Our 'New Mother' – Benze with Bodil and Barselaj

I'm going to sleep, I'm closing my eyes tight'. Then we will say the Lord's Prayer together. At seven o'clock, you'll be woken up; you'll be accompanied to school tomorrow, so that you can learn where the Danish school is located, and how to get there. Principal Binzer will be waiting for us at eight o'clock, in order to welcome us, and after that, we'll be photographed.' She concludes by saying, 'So you'll have to get going.' 'Hold on a minute', I think. 'That's a lot of things to learn and remember'.

When we finally get into bed, we whisper 'goodnight' to one another. The bed is nice and big. I have a hard time falling asleep – what a day it's been! At the beginning of the day, I was looking forward to seeing my mother and my siblings. Now, it turns out that I can't even talk to them and that I'll have to stay here at the orphanage, instead. When can I go back home and live with my mother? The questions spin around my head. Agnethe is lying in the bed next to me; I can hear that she's been sniffling, and we can hear Little Karl crying, '*Aaqa, Aaqa*'. His crying is infectious. The tears flow down my face, and I crawl right down under the bedclothes, silently sobbing in despair. I am deeply unhappy, and I whisper, 'Mom, Mom, please come and get me'. I hear a movement, and in the dark, I can see Ane Sofie sneaking out of our room and walking to the room where her little brother is.

In the morning, we are awakened by a knock on the door. 'Good morning! Let's get up now'. This is followed immediately by, 'Where is Ane Sofie?' We

FIGURE 9 Miss Blom

wake up immediately, and look over at her bed. It's empty. Then, from the next room, we hear, 'What are you doing in your little brother's bed, Ane Sofie? You have to sleep in your own bed!' Little Karl is crying and trying to hold on to his big sister. Ane Sofie wriggles out of Karl's bed and walks over to fetch her toothbrush. I go out to pee, closing the cubicle door with the small swivel lock, but the toilet bowls are high up, and they smell, so I finish as quickly as I can. The toilet paper is thin and stiff, and I have to fold it between my fingers a few times before it gets soft enough to use. I can't hold my breath for that long, so I rush out of the cubicle and into the bathroom. Fortunately, there's a washbasin free.

When we sit down to dinner in the dining room, we have to hold our tablespoons out horizontally, in our right hands. Then a *kiffak* comes in and pours cod liver oil onto our tablespoons, which we have to swallow, quickly. What a taste! So when Miss Bengtzen gives us permission to start eating, we waste no time in getting the first spoonfuls of food into our mouths. Then a second *kiffak* comes in and gives us a vitamin C pill each. There is froth in the milk, and it is lumpy, and we ask why this is. Miss Bengtzen explains that it's been made from dried milk; the powder is whipped up with cold water, and it is not always easy to do this. When we girls start working in the kitchen, we will have to try to whip the milk ourselves.

School and Weekdays

We are woken up at seven o'clock, by one of the adults knocking on the door. The door opens, and we hear, 'Good morning! Let's get up now'. It's difficult to get up sometimes, but it's not possible to sleep any longer, because of the noise from the bathrooms, and people walking up and down the stairs. So I might as well get going and queue for the bathroom. Every morning, I think to myself, 'I hope my hair isn't too messy today'. I have to make my bed, and get my school bag ready, before going down to the dining room. There, we have to sit to attention, with our backs straight, and our hands at the edge of the table. We're not permitted to sit with our hands in our laps – 'Because you've just washed your hands, and you never know what people might be fiddling with', we're told. After breakfast, we pick up our packed lunches, which consist of two thick slices of rye bread, one with liver paste, and the other with salami, wrapped in baking parchment. The cook, Karen, passes us our packed lunches through the serving hatch. After that, we have to go and get dressed for the day. When we are ready, we stand in line, two by two, and hand in hand, with the youngest children at the front.

At first, an adult accompanies us to school, but soon we are allowed to go by ourselves. Mostly, this goes well; but when the lakes that we have to walk past freeze over, it's hard to hurry straight to school. The big boys, Eli and Ole, ask, 'Who dares go out on the new ice in the middle of the lake?' Of course, Barselaj does. It's fascinating to watch him, because we can see the new ice bending beneath his feet, and hear it start to crack. Suddenly, Barselaj falls through, but

FIGURE 10 The dining room

luckily, only up to his chest. The new ice looks like shards of glass, which break into a thousand pieces when he goes through. Barselaj has some difficulty getting out, and he is shivering with cold, and his teeth are chattering, as the other boys help him. Eli goes home with him, so that Barselaj can change into some dry clothes. The rest of us hurry off to school, but we're late when we get there. At dinner, we get a talking-to. We must go directly to and from school, and not go off and play along the way. 'Do you realise that if you're late for school three times, you'll get detention?' Everyone falls silent, as this news gives us all something to think about. However, despite the trouble, we still get our evening story. Miss Blom is reading us 'Uncle Tom's Cabin', which is exciting. When the tables have been cleared, and the washing-up done, she likes to sit at the end of the table and start reading aloud. Clumsily, we try to position ourselves as near as we can to her, sitting on the bench or on a chair at the dining table, struggling amongst ourselves to get closest.

After this, we can decide for ourselves what we want to do and where we want to be. We might be in our own rooms, or with others in theirs; or in the dining room, where we can play games; or, we can go out and play until 7.30 pm. At that time, we have to start getting ready for bed. At exactly eight o'clock, we have to be in our nightwear, kneeling in front of our beds, with our hands folded, and saying the Lord's Prayer ('Our Father') in unison. Then Benze goes around saying, 'Good night, sleep well.' I often think about my father. Why did he have to die, too? My tears fall onto my pillow. Why did I have to stay at the orphanage, when I'd thought that I'd be going home to my mother, and my big

FIGURE 11 Barselaj and Miss Blom

sister and my little brother? My tears are flowing, and the edges of my sheets are completely soaked now. I don't want the others to hear me sob, so I crawl right under the bedclothes and think, 'Why? Why?' The next morning, my eyes are swollen from crying, so I quickly splash cold water onto my face. My hair is messy again, and I wet the comb before I use it, but my hair sticks up anyway. Then I stand by the stool next to my bed and put on the clothes that I folded neatly the previous evening. Then it's time for breakfast, and the cod liver oil. Just the thought of it pains me. I can't believe that we have to use the same spoons for taking our cod liver oil that we do for eating our porridge. It's almost impossible to get the taste of cod liver oil off the spoon, but I try to eat the porridge quickly, and it's lucky that it tastes so good.

One morning over breakfast, Benze says that the people in Big Kristine's will have to help find a present for her from the gift room, because her birthday is on the 23rd. It's nearly autumn now, when we'll start at our new Danish school. The weather is good; whilst it's not windy, it's feeling colder already. Our new Danish teacher's name is Miss Else Petersen, and we have to address her as 'Petter'. She's not young, and she has a big head, with her hair pulled into a bun at the top of her neck; she wears big glasses, has big front teeth, and a hefty bosom. She's friends with Benze. Our first reading book is called '*Ida and Ole*', but first, we must learn to write out the alphabet. Petter writes each letter in turn on the blackboard; she writes neatly, and it looks like a difficult task. We each get a practice booklet, and when we are skilled enough, we get an ink-well and pen. Then we have to make a real effort to write properly. Not only do the letters themselves have to be well-formed, but they also have to run beautifully together. We get some good news one day – all schoolchildren now have to learn something called 'form-writing'.

This means that we don't have to use italics anymore, because that's considered old-fashioned now. Luckily for us, it's easier to use 'form writing', because the letters look just like print letters.

In the middle of the day, there is a long lunch break. 'Get your lunches out now', the teacher says. It is always interesting to watch the Danish children unpacking their lunches. Their lunches are packed in nice boxes, and inside those, their food is wrapped in baking parchment, and there is paper in-between the different food items – no items touch each other. They also have lots of different types of food – along with liver paste, there might be beets, pickled cucumbers, or egg yolks with chives, and homemade mayonnaise. Some of them even bring in chocolate. Yet some of them say that they don't like their food. I think that they should be ashamed of themselves. Their mothers have been shopping at the new butcher's, which our classmate John Petersen's parents have opened. It is located directly opposite the Duplex, my father's old workplace on Skibshavnsvej, and I have been in there myself, as my Aunt Kristine works there. She sometimes passes me a bag of pork over the counter to me, saying, '*Iggu*' ('You're so sweet').

The boys are the first to ask the Danish school students, 'Can I have the food that you don't want?' I look on, enviously, but I don't dare ask. Only much later on, after becoming good friends with some of the Danish girls, do they ask me if I want one of their sandwiches. Yes, I'd love one. Their rye bread slices are so thin that the sandwiches are difficult to hold, but they certainly taste good. One of the first friends who invites me over to play at her house is Lone Berthelsen. She lives in Rævedalen. Her mother is Danish, and her father is a Greenlander and a school inspector, but she looks like a typical Danish girl. We are told that we shouldn't invite our classmates home at first – not until we have got to know Godthåb better. On some Sundays, the staff members take us out and show us where our Danish friends live. One Sunday after lunch we go out to Østerbro, where a lot of Danes live. Benze was a nurse at Godthåb Hospital between August 1946 and April 1949, and again, from November 1949 until August 1952. Therefore, she already knows most of the Danes in Godthåb, who lives in the different houses, what their names are, and what jobs the men in the houses do. At that point in time, all of the Danish women were at home.

The Danes' houses are wooden and painted in different colours – red, yellow, green, brown, and grey. Some are bigger than others, but they all have conservatories extending from the living rooms, with flowers or tomatoes. One of my classmates lives in a green house. I think she is a really spoiled Dane, because she has a dark green wool coat, with a leopard fur collar, and she goes out playing in it. One day when we pass by her house, she is out skipping in the coat. When we get back from school, we have to change into our worn-out clothes.

To get to Skibshavnsvej, we have to pass the Greenlanders' cemetery. It gives me a pang in the heart, because my father is buried there. I remember his funeral with sadness, and I still can't believe that his life is over. I look over at the cemetery. Next to it, something's being built; it's going to be a sanatorium for tuberculosis patients, and the adults tell us that it'll be named Queen Ingrid's Sanatorium. We also pass Mrs Slot, who cuts our hair and lives in a red house on Skibshavnsvej. She has just had a baby, so the following Sunday we visited her, and had cocoa and buns with Benze. On the way out to the harbour, we pass a yellow house, where I spot one of the big boys from the Danish school. He's looking down at us from his room; I think he's really nice, so I'm startled when we make eye contact. From his house you can see the Sailor's Home, where one of Eva and Marie's classmates, Margaret Mikkelsen, lives. From her home, you can see the harbour that we sailed into, when we came home from Denmark.

One day, we go down towards the KGH, and I recognise a green house, where one of my old playmates lives. Before my father died, I used to play there with Hanne Chemnitz. We pass by the US Consulate, which is a large brown building. The county council hall is nearby, which is a red building, and from there I can see my father's old workplace – the telegraph station – and to the right of that is our house. My heart starts beating faster as I see my mother and my siblings, when we pass by. Right next to our old neighbours' house – Uvdloriánguaq and Henriette Kristiansen – my big sister jumps out onto the road and stands right in front of me. We give each other big smiles. She has long, thick, black braids which I think are beautiful. I want to tell her this, but we're not permitted to talk with one another. I quickly touch her braids, and wave to her, before we move on. I really want to stay there, playing with my big sister. She gazes after us, looking completely lost.

We are going out into the middle of the marsh, where there is a big red house. To get there, we have to cross a wooden bridge, and we visit Kirsten, from the Danish school. Her parents are Mr and Mrs Helge Andersen. The mother always has perfect hair, with curls like those of an actress, and she has very red lips and nails. They have plenty of space. We call in, and have some cocoa and buns, before we continue along Skibshavnsvej, past the new Winstedt radio shop. We look at the display in the shop window, which is filled with small transistor radios, and we discuss whether or not you can really hear the news on such small radios. Our old neighbours, Telef Lynge and his wife, live next door. I recall playing with their daughter, Juliane, when my father was still alive. After we pass their house, we get to that of the police chief, Vesterbirk, who lives in a blue 'Danish house' – the emigrants' houses are easily recognisable. Later on, their son Lars will join my class. This is an exciting place, because it is a police house, with a prison in the basement. The next house to that is a big, green house, which belongs to the dentist, Dr Albrechtsen. Benze knows that family, too. A little further on, there is a strange, long green house; it looks like a giant half-barrel, and it is the assembly house. To the right of that, there is a small, brown house, which looks like a small bar, but is Godthåb's new grocery shop. At the end of the marsh, there is a small, green baroque house, which is Ole's Department Store, where you can buy – amongst other things – shiny pictures and dolls that you can dress up.

Out towards Skibshavnsvej, there is a red bungalow, which has a dentist's on the first floor. Next to it is a restaurant called 'Kristinemut'. There is always a terrible noise going on over there, so we scarcely dare to even look in that direction – just imagine if we saw a drunkard! Some of the boys say that they have seen an old man there who was in this condition – he was a thin, dark, and wrinkled, wore dirty clothes, and he could hardly even walk. He was staggering away, with a crowd of children in pursuit, who were shouting at him. He had turned around, waving his arms in the air, and had tried to run after them. The children ran away, screaming, until he turned around again and staggered on his way. The same thing happened over and over again, all the way along Skibshavnsvej. I felt callous towards him, and pity for him, at the same time.

Finally, we reach Rævedalen. There are many large houses there, where more of the emigrants live. The houses have basements and first floors. My classmate Lone lives in one of the fancy houses. Now I know where she lives – in the red house. A little further on are several blue, terraced houses. First and foremost, the houses are for unmarried emigrant women, and therefore they are known collectively as 'Trussely'.[4] We giggle and look furtively at one another, when we hear about this. At the very bottom of the road, there is a fantastic view of the police station in Old Godthåb. The chief of police, Simony, lives in the former mission church. He is also one of Benze's friends, and we are invited in for juice and buns. Their *kiffak* comes over to me and gives me a gentle hug. Her name is Birgitte, and she is one of my mother's friends. She's a lovely lady, who I know her well, and even though we can't talk much with one another, it's wonderful to feel the warmth radiating from her.

On the way back, we cross behind the telegraph station. It gives me a start, because there in front of me is the English cottage, with the small, white observation box, two metres above the ground, and its small white staircase. When my father was alive, he used to send up the white weather balloons from this very spot. The English cottage is a mini weather station and has a thermometer, hygrometer and barometer. Every four hours, the balloons are sent up; half a bottle of hydrogen is used in each one. Because of the risk of sparks from the light switch in the dry air, the English cottage's light switch is in another room, at a distance from the hydrogen bottles. My father used to measure the direction and speed of the wind using these balloons. A minute after being sent up, the balloon would be at an altitude of 300 metres and after a minute and a half, at 450 metres. He could see the way in which the balloon was drifting and gauge the height of the clouds. If the direction of the balloon's drift changed, this of course showed that the wind direction had changed. The weather observations were first called into Ammassalik, and from there to France, where the International Weather Forecast Centre was located. As a telegraphist, his duties also included sending out telegrams, and copying and distributing the Godthåb newspaper. This newspaper was written on two pages and was distributed daily to all of the Danish households. A single-page edition, written in Greenlandic, was hung up in the colony harbour, where the locals could stand and read the latest news.

In addition to this daily newspaper, from 1952, the four-page weekly newspaper, *Kamikken*, appeared, which was written in both Danish and Greenlandic.

Eli's linoleum print of 'The Ugly Duckling', and Ole's linoleum print of Hans Christian Andersen, appeared together in that year's issue number 20. My linoleum print was also published in *Kamikken*. Other than that, the news that was reported in *Kamikken* was mostly about who had arrived on the ship from Denmark. To give an example, here is an extract from *Kammiken* on Tuesday, June 10, 1952:

> PASSENGERS TO GODTHAAB[5] WITH QUEEN ALEXANDRINE III: National physician Mogens Fog-Poulsen, assistant Egon Nielsen, cooper[6] Kønig, electrician Jørgensen, nurse Ingrid Hansen, forty-eight men and two women for Asmussen and Weber, eight men for Højgaard & Schultz,[7] engineer Magnussen, diver Carl Hansen, ophthalmologist Fabricius Jensen, pastor Nygaard, carpenter Poul Petersen, principal Binzer and his wife and four children, house assistant Amalie Geisler, Master[8] Meldgaard, Georg Nellemann, archaeologist Claude Desgoffe, Master's student Inge Parbøl, parish priest William Larsen, Asia Simony, economist Grethe Hansen, and six conscripts.

We walk back up the slope of Inspektørbakken. One of the *kiffaks* tells me that my mother's aunt lives in one of the larger of the red houses. We can see our way home from that point, and we race the final stretch, all trying to be the first one back. We go in through the basement door, and up into the kitchen, to see what we're having for dinner. It's cod – again. There are several large cod out on the table. 'What are we having for dessert?' we ask Karen, our cook. 'Sago soup', she replies. 'Oh, marble soup!', we reply, because we know that there are round balls in the soup, which look like eyes, but are red, and sweet-tasting. Then we run towards our rooms to play, but Miss Blom takes us aside, checking to see if we have any homework to do. I have some lines of form writing that I need to practice, so I join the other people who are sitting in the dining room, and we get on with our homework. After that, we can go out playing. The sun is shining, and the weather is wonderful. The first snow has settled on the hills, and we sit outside at the back of the orphanage, and playing with our dolls. I am playing with my china doll, Tove – the one I got from my foster mother, Ingrid (Figure 12).

'Look, we've found gold!' shouts Ole shouts. We put whatever we're holding down, and run over to look – and, yes! – in the black, porous stones which Ole has found, something golden is glittering in the sunlight. 'Run in, and show it to Benze!' we urge him. We burst into Benze's apartment, full of excitement, looking expectantly at her facial expression as she examines the stones. She goes over to stand in the sunlight, and we circle around her, awaiting her judgment. She turns stones over and over, and eventually, she says, 'Well, it might look like gold glittering there, but there's no gold in Greenland'. We are hugely disappointed. We go out playing again, until the bell rings for dinner. Then we have to wash our hands, move our chairs up to the table, place our hands on top, and straighten our backs, waiting to be invited to start eating. We always have to wait until the

FIGURE 12 Homework in the dining room

last person has sat down, and often this is Benze, because she is always so busy. That day, we had fishcakes and *rødgrød*[9] for dessert. Then, here they come again – the little Greenlandic children from the town are standing on the terrace, shading their eyes with their hands, and staring at us, as if we were monkeys in cages. 'Benze, they're here again!' we shout, and the *kiffak*, Bodil, opens the French doors, and chases them away, using a mass of Greenlandic words. 'We must have some curtains made for the French doors', says Benze. 'I'll tell Gertrud, when she comes in tomorrow'.

Our seamstress, Gertrud, is already very busy, sewing burgundy checkered woollen trousers, and school anoraks, for all of us. The next day, Benze goes down to the KGH to buy the materials. We girls ask her, 'Can I come along?' I often go along on such trips, because I finish school early. We skip and run down the road, where a lot of construction is going on. At the end of Børnehjemsvej, they are building a red house, and all around it, there are piles of building materials. Over to the left of that, they are putting foundations into the rocks to build the new radio house. There's a terrible howling sound every time the rocks are split, which is a warning for people to not come too close. Then there's an enormous noise, followed by a cloud of dust which is so thick that you can't even see your own hand in front of your face. We can see all of these goings-on from the windows of our dining room.

We walk out of HJ Rinksvej, past the US Consulate, and just before we reach the KGH, I spot my mother. A man behind her is pulling a cart with a white coffin on top. I run over to my mother, feeling embarrassed, because she looks so distraught. I ask her what she's doing. One of the *kiffaks* translates what I've said, but my mother puts a finger to her lips, and strokes me with the back of her hand. Afterwards, the *kiffak* tells me that my grandfather is dead. I feel hurt, and I sigh deeply, wondering why no-one has let me know about it. Now I'll never see him again. After this shock, I can't really remember getting to the KGH, but I do recall seeing the endless shelves of rolled-up fabric, arranged in long aisles. Gertrud chooses some transparent white fabric, and we go to the hardware store to pick up some white plastic cords, and four hooks, to hang the curtains from. The next day at dinner, the curious children can no longer see us. It is a blessing; from that day on, we no longer feel watched.

Sundays and Birthdays

The door is opened with a 'Good morning! Let's get up now'. Half asleep I think, 'What day is it? Oh, no, it's Sunday.' We have to go to church for the Danish service at 10 a.m. I go over to the window and draw the curtains, to see what the weather is like. Luckily, the sun is shining. There's a shout from the hallway: 'Who wants to go out and raise the flag?' This job has to be done by 8:00 a.m. I quickly wash my face, brush my teeth, and jump into my clothes. Around the flagpole, there's about half a dozen of us who are ready to help with the hoisting. The Dannebrog is flown on Sundays, and I'd rather join in the raising of that than the Red Cross flag, which is flown on every other day of the week. First, the flag must be carefully unfolded; then the proper attachments must be made at the top and bottom corners. That day, I am allowed to hold the flag so that it does not drag on the ground. Benze is in her Sunday clothes and slowly raises the Dannebrog to the top of the pole, whilst we sing, '*I Østen stigen solen opp*'.[10]

Over breakfast, we whisper our complaints to one another about having to go to church again, but when Benze looks over at us, we quieten down, and we smile politely instead. When everyone is ready, we head down to the church. As usual, there is enough room in the church for the Danish church service, and we nod at the other churchgoers. I sit on the third row with Agnethe, Bodil, and some of the other girls. We sit swinging our legs, jostling one other and giggling, until some of the adults whisper at us to be quiet, because the service is starting. We raise our heads as high as we can; our noses just about reach the top of the backs of the pews. We sing Danish hymns, and then the priest gives his sermon. It takes a long time, which we spend yawning and looking at the images of Jesus on the cross, and the angels, and the nave – and sometimes, we gaze longingly out of the tall windows. At last, the service is over. Afterwards, we sometimes go with Benze to have a 'drop' of coffee – either at the parson's house, which is next to the church, or at the Governor's place, which is a long, grey house with a big garden, which stretches all the way down to the harbour.

'There's a whale!' The rumour starts down at the harbour, and it quickly reaches us at the orphanage. We set off, with some of the *kiffaks* and the boys, down to the bay by the hospital. Whales are huge; more than twenty people could stand around one, and there'd still be room to spare. Some men, who are wearing big rubber boots, are standing on top of the still-steaming whale and cutting big lumps of flesh from it. Eventually, you can see its gigantic intestines. We have brought a zinc tub with us, so that the boys can carry some of the meat and whaleskin home. It seems that there's enough meat to feed the whole town. Some of the men cut themselves pieces of meat directly from the whale. They take a lump in their hand, cut it into smaller pieces, put it straight into their mouths, and begin chewing right away. Our mouths are watering, because we want to do that too, but we'll have to wait until we get home to the kitchen. First, the meat will have to be rinsed, then cut into very small pieces, and then we'll each get a little bit of it on a small plate. The next time we walk by the front of the hospital, only the gigantic skeleton of the whale is left.

Sometimes we go out with Miss Blom. One day, we walk past the red barracks near the orphanage in a small valley. Soon, we'll be enjoying this view for the last time; Miss Blom tells us that it's going to be blocked off by dams at each end and filled with water to create a large reservoir, so that the whole of Godthåb can be supplied. Because there is no natural flow of water here, the reservoir will be refilled each summer via a pipeline that has been extended to the lakes behind Little Malene, several kilometres away. It is strange to thinking that soon we'll be drinking water from the place where we are now standing. A little further up the mountain, there is a steep precipice. We look over at the drop, but we know not to go all the way up to the edge, because that is the exact spot where a woman fell to her death from. We shudder, because we've heard that story, and it had been just as rainy that day. I remember that it was on King Håkon's birthday. Benze was a true royalist, so we knew all the names and birthdays of the members of the Nordic royal families.

It's interesting to watch the construction of the dams. Right behind the orphanage, they set up a stone crusher, which was sometimes really noisy, and there are piles of stones of varying sizes. It is mostly the boys who watch the tradesmen work. They live in barracks, with their families, and we become good friends with their children. We can visit them when we like, and their children can visit us, too. One day, we all went to a birthday party in the barracks – they were all really nice. One autumn day, the boys come home and tell us that one of the tradesmen has caught a fox. We go over to see if it is true, and it is. They have made a cage for it in the mountains behind the barracks. It's interesting to kneel down by the cage and watch the fox. I remember that this was around the time of Aaron's eighth birthday, October 25. The boys from his class were allowed to go into the gift room and choose his present; we girls felt jealous, because it was exciting to be one of those doing the choosing. There were lots of shelves, with toys for boys and girls on most, arranged in rows – a real wonderland. Aron gets a toy truck. Later on, Little Kristine told me that she and some of the other children had found a secret entrance into the gift room. You could crawl in through a hole in the wall of the hospital room and get in that way. It was nice to have the time to look at all of the presents – I hadn't previously why some of the others could choose a gift so quickly! On children's birthdays, the table would be covered with a white damask cloth, and cake plates and coffee cups laid on top. There were flags on the table, and we got buns, cakes, and hot cocoa.

Look at My Angel!

The first snow comes, and it keeps on snowing. It is great to experience a real Greenlandic winter again. There is so much snow that a nearby gorge fills completely with snow. We have fun, taking turns at jumping into it. Sometimes, one's boots get stuck, but we usually manage to get them back again. Ane Sofie takes a big run-up and plunges into the soft snow. Suddenly, we can only see the top of her hat. Luckily, there's enough of us there that someone can run home to fetch an adult, whilst the rest of us try to help Ane Sofie. Miss Ingrid comes along and sees what's happening, and she asks one of the boys to run home to get our janitor – he'll need to bring a bucket. We have dug Ane Sofie out as far as her shoulders, but now the surrounding snow has frozen solid. Jens arrives and starts gently digging. Little Karl has also got here, and he's crying, calling 'Aaqa, Aaqa'. We comfort him as best we can, and at last, Ane Sofie is free. That day, we learnt respect for the deep snow in the mountain gorges.

In the early days at my family's home, we had neither a freezer, nor a refrigerator. However, Benze bought things in bulk, as she was extremely frugal. A lot of rosefish were available in the town; Benze had heard about this, so she immediately sent a *kiffak* and Jens down to buy a load of them. We had thought that it was enough having boiled cod and pickled-fish soup. It was interesting that the rosefish were really red,[11] but their big blue eyes looked ugly. The boys helped to bury the excess rosefish in the snowdrifts around the orphanage. A week later, when we had rosefish again, they didn't taste as good. It had been raining for a couple of days in the meantime. We sat down to the rosefish reluctantly, but were told to eat up. We were quiet as we tried to eat the rosefish, and Benze told us that we'd have to eat anything we left the following day – 'Then you'll learn to eat when you're told'. The next day, there was a lovely smell of roast grouse, and we thought that Benze been joking, and that we wouldn't have to eat the leftover rosefish after all. But when we were called in for dinner, we discovered to our horror that we were having the rosefish. It smelled rotten, but we had to eat it anyway. I got physically sick. We children were in a very bad mood, especially when we saw that it was only the adults who were having the grouse, which Miss Ingrid had got from her parents in Kapisillit. We glanced at the delicious grouse, feeling very small and insignificant. We could see that the *kiffaks* felt sorry for us. Most of us ate only the potatoes and curry sauce. From that day on, I have not been able to stomach rosefish.

It was a great day when our newly made winter clothes were finished. Gertrud had been busy. We girls got long, woolly plaid trousers, in different shades of burgundy, and a burgundy-coloured anorak; the boys got grey tweed plus-fours. It was also time to change into our hand-knitted woollen sweaters and our thick flannel underwear with the elasticated legs. It felt heavy, compared to our cotton underwear. Occasionally, the water truck came up the road to fill the water tank; on other days, the tanker that filled the oil tank in the basement would arrive. The boys would be sent out to shovel snow, so that the water or oil hoses could be connected to the building. We girls would make socks. We liked sitting at the big dining room windows with Miss Blom in the dining room. We each got needles and learnt how to knit. We had to start by making big stitches around a frame, avoiding making knots in our work, because these would hurt the wearer when walking. Then we had to put in some vertical stitches, then some horizontally, and finally, work in-between the vertical ones (Figure 13). Then we'd have a finished sock. Sometimes, we chatted or sang when we worked. Once a month, we girls helped to change the bedding. This was easiest in winter, because during the summer, the thick woollen mattresses had to be beaten out with a carpet beater. We'd have to wait for the boys to do this before we could change the bedding. We went around like little *kiffaks*; those who worked in the kitchens helped change the bedding.

The orphanage was a brand new building and was solidly built. Inside the house, we could hardly even hear the harsher blizzards. When the blizzards were at their most severe, we couldn't go out playing, even though we wanted to. But once they had subsided, we would, especially during those first winters, go out and enjoy our lovely Greenland again. It was great to just feel the snow and the cold, and to jump and play in the snowdrifts; or lie flat on your back on the newly fallen snow, looking up at the snowflakes falling on your face. When lying like that, we'd move our hands from our hips to over our heads, and at the same time, our legs from side to side. When we stood up again, there would be an imprint in the snow, which looked like an angel with big wings. 'Look at my angel! Look at mine!' we'd shout to one another, and we would dance around, enjoying the winter to the fullest (Figure 14).

We grew up in the new, modern Greenland. The electricity supply from the first power plants was not always stable, especially in winter. One evening after dinner, we were sitting around the table with Miss Blom, who was reading 'Lille sorte sambo'[12] to us. Suddenly, it got completely dark and screamed in terror. Miss Blom calmed us down, and Benze fumbled her way into her apartment. A little while later, Benze came back; we could see her by the light of the green kerosene lamp that she was holding. She set the lamp down on the dining room table, and we wondered aloud what had happened. 'It's that electrician, who kisses his wife', replied Benze. We looked at each other and giggled, and then looked back at Benze, wondering how she knew about this. Miss Blom continued her reading, by the light of the kerosene lamp, and we thought that this was even nicer. When the power came back on, we had to get ready for bed. Benze went around

FIGURE 13 Me (left), knitting

the rooms, to see if we were kneeling by our beds, singing '*Jeg er trært go gå til ro*',[13] and saying the Lord's Prayer together.

One day we were in the dining room, with the new coloured pencils and drawing paper that we'd been given. We were watching out in case Benze suddenly came in, because it was her birthday on November 27, and we were drawing and colouring pictures for her. If she came in, we'd hide the drawings underneath our school books, because we usually sat in there doing our homework. The cooks were busy baking muffins and pies and also making a fine dinner, because that evening Benze's Danish friends would be visiting. The night before, it had been 'the Lord's own weather', as Benze put it; one might think that God Himself had painted the orphanage white, because the red woodwork of the building was completely covered with snow. Miss Blom got a tray ready, with a cup of coffee, a small Dannebrog, a candle, and a piece of French bread. Then we tiptoed into Benze's bedroom and woke her up with a birthday song. 'Today is Benze's birthday!' we sang out. I was surprised when I saw her, because her thin, blonde-brown hair was hanging down loosely, and looked ugly; also, she wasn't wearing her glasses. She looked completely different, but also happy, and surprised. She admired all of our pictures; I can't remember what other gifts she got. When she got up later, we saw that her hair was neatly drawn into its

FIGURE 14 Playing in the snow

usual bun. When school was over that day, we had hot cocoa and muffins, which was a nice surprise; we also had a good dinner later on. That evening, Benze's Danish friends visited. After their coffee, which was served on Benze's three-decked silver cake stand, they were given birthday cake and cognac, served in her crystal glasses with the black bases. We looked closely at the guests, with their beautiful clothes, especially the women, with their striking red lipstick and red nail polish. We fell asleep to the smell of coffee and the sound of the blizzard outside in the dark. In the *Godthåb Avis* [Godthåb News], they wrote about how bad the weather had been.

Some of the boys, who were out in the long hallway, were shouting out, 'New sledges!'. We went out there to look, and they were quite right – there were four new sledges for us to share. We got dressed quickly and went out over the football field, up the hill, and then we could start sledding back down. Before the snow was properly packed down, the first few descents were slow, but it got faster and faster. Best of all was when there were two people on the sledge at the same time, when we would scream with delight. Eventually, we could slide down almost to the very end of the football field. Although it had become dark by then, we could see quite easily in the light from the moon and the Northern Lights. With our cheeks now glowing, we put the new sledges away under the stairs to the basement, just before dinner.

We had been told that the last ship to Denmark before Christmas would soon be setting off. It was late November, so after we had done our homework, we all had to gather in the dining room and write Christmas cards to our foster families in Denmark. We were each given a postcard, and we had to bring our pencil cases in – there were erasers on the table – and as a surprise, there was a bottle of soda for each of us. I enjoyed it, but for most of the boys, it was pure torment, apart from the soda. They drank theirs very quickly, but Agnethe and I tried to save every precious drop, so we had our sodas for a long time. The next people to have their birthdays were Eli, on December 6, and Ole, on December 9. The older boys went into the gift room to choose the presents, and it was difficult for us girls not to be with them. We waited outside in anticipation, but the presents were already wrapped when they brought them out. Eli and Ole were allowed to bring a couple of the boys from their classes at school home. We were always excited to sit down at the table for birthdays, but we had to wait until the last kids had arrived back from school. As soon as we saw them coming over the mountains we shouted, 'They're coming!'. Then we could go and wash our hands and enjoy the freshly baked muffins, hot cocoa, and birthday cake.

In the middle of November, there was a good deal of activity. There were buns and cookies were to be baked, because Benze had invited our mothers for coffee on one of the Sundays. Only we six children from Godthåb could have visitors. We could hardly wait; we were glowing with joy, and we couldn't talk about anything else. But then I caught sight of Little Kristine and Ane Sofie, sitting there with long faces. 'What's the matter?' I asked. 'Well, you can see your mother, but some of us'll never get to do that!'. Anso and Little Karl's mother was dead, and Kristine's mother lived far away, up in northern Greenland. This dampened my joy, and when my mother finally arrived, something was burning inside me. I had a huge desire to say to her, 'Won't you take me home after the coffee?' But I didn't dare say it. If Benze had heard me, she would probably think that I was being naughty and send me to a completely different city. If she did, I wouldn't see my mother again. I also didn't know if my mother would have understood my question. So I just walked over, and leaned in closely towards her, holding her hands as I stared intently at her, with a serious expression.

One morning in December, we woke up to the sound of the wind blowing, and the weather seemed warmer. We got out of bed quickly and looked out of the window. The whole of Godthåb was covered with ice. Everything – the mountains, the roads, the houses, and the snow ploughs – was all covered with ice. It was as if someone had turned Godthåb into 'Ice-håb'. Some of us quickly put on our jackets and boots, so we could go out and slide on the ice. It was fun; we couldn't even walk two steps without falling over. We heard people shouting, 'Come in, breakfast's ready!' When we found out that the doctors had called Benze to tell her that the school was closed, we finished our food very quickly. We were filled with joy when we heard that the school would be closed for several days. It was closed because of polio, but we didn't know then what polio was. During the winter days, the electric lighting failed quite often, and we often used the kerosene lamps.

Christmas Preparations

Benze had stored the blackberry leaves, from the summer we had arrived, in the basement. She put them underneath the spruce branches that she used to decorate our large Advent wreath. It was fun following the Advent stages. Four big white candles were placed on the wreath and big loops of red ribbon. Then it was hung from the ceiling, in front of the large windows, the day before the first Sunday in Advent. On the Sunday morning, when everyone had sat down, the first of the candles was lit. During the evening meal, we were told about Jesus' birth, and we had lots of questions. 'What does a crib look like?' 'Why are we still celebrating his birthday?' We ended up having to borrow some books about Christmas from the library. We didn't have to wait for people at breakfast in the month of December – everyone was already sitting neatly at the table when the bell rang. We were so excited. An Advent calendar was hung up on the dining room wall. We were each allowed to open a door in turn, beginning with the youngest of us. I was aged somewhere in the middle, so I opened the door on December 8. It was a picture of the Little Match Girl.[14] She looked so sweet. There was always a rush of people trying to get up on the bench and look at the Advent calendar pictures – we almost had to press our little noses up against the calendar, because the lighting from the ceiling lamps was so poor. A little further along the wall was our big Christmas tree. We had to take it in very carefully, so that it would fit.

It was nice, snowy weather, and the bottom steps up to the classrooms were covered in snow. As we entered, a large green Christmas tree was drawn on the blackboard, as well as hearts, wooden cabins, and candles. Every day, someone would get to light the candle. We took turns doing this according to our surnames in the register, so it took a while before my turn came around. One day, we got a new reading book, entitled '*Ida and Ole*'. It was similar to my book, '*Frederik*'. Else Petersen clapped her hands and said, 'Now, you need to hear about a new scheme that has come in. From today, you'll have to queue up in front of the entrance to the school yard, where a couple of ladies will hand out portions of *råkost*[15] to every schoolchild. This is what the National Council has decided, because children in Greenland must be fit and healthy'. I thought that this sounded nice – because now we'd have both our packed lunches, and a portion of *råkost* to eat – but the queuing would take up some of the lunch break, and there was new snow today. For me, it was all about getting outdoors as quickly as possible.

After school was over, we did some model-making. It was fun; we drew and cut out two A4-sized cardboard Christmas trees and attached them together, using the set of notches that we had made. I wanted to bring my cardboard Christmas tree home to my Mom, when I went home to her for Christmas. I told Benze about this when we got home from school. She was completely silent; I looked up at her, and I saw her serious expression as she said, 'No, none of you children from Godthåb should hope to spend Christmas at your mother's houses. Now you live here, at the orphanage – that's where your home is, and it's where you'll have Christmas. We have already bought everything we need for Christmas. We must have a real Danish Christmas, with everything that comes with it. You can look forward to that.' Agnethe was standing next to me. We were completely mute; we quickly looked at one another and ran back to our bedroom. We closed the door and hugged one another, and then our tears started flowing. We understood nothing, and we said even less than that. But inside, our hearts were also crying.

After dinner, we read '*Peters jul*'.[16] Some evenings later, once we'd found out about everything that was to follow, we were looking forward to Christmas. All four of us would whisper in the room, talking of the days leading up to Christmas Eve. We had seen Benze and Miss Blom go secretly into the gift room. Later on, when we heard Benze say 'Good night' to Miss Blom, Agnethe got her cardboard Christmas tree out, and the four of us danced, on our tiptoes, around the tiny tree, quietly singing. Eventually, though, we forgot about singing in whispers; and suddenly, there was Benze, standing in the doorway, smiling and inquiring, 'What are you doing?' Without answering her, we jumped into our beds. The next night, we did the same thing; again, Benze came, and again, we jumped into our beds. We had listened out for her approach, and danced faster and faster around the tree, singing louder and louder. When the door opened, and we had jumped back into bed, she lifted our quilts, and gave us each a gentle pat on the back. Then she put our quilts back over us, saying, 'You can get yourselves back to sleep! Good night!'

One day, Agnethe and I were allowed to help Karen in the kitchen, baking iced buns. The big mixer was up and running. First of all, using a measuring cup, Karen poured the flour in, from the sack that was on the table. Then, we put in the brown sugar, and we were each allowed a teaspoon of the sugar. Then we put in the eggs and the spices; then Karen put the big kneading hook onto the mixer and switched it on. We were standing close by, but we jumped aside at the noise that the mixer made when it started. We went up to it again, to see what was going on. When the dough was firm, and light brown, Karen stopped the mixer. Then it was time to make the icing, but Karen had left the food dye in the cellar. I got the key, slipped down to the cellar, and found the little glass bottles that we needed. There was a red bottle, a yellow one and a green one. I locked the door behind me, and I started up the stairs with the three bottles in one hand, and the key in the other. When I got the top step, I tripped over my own feet and dropped one of the bottles, which smashed, and cut the ring finger of my left hand deeply. 'Damn!' I shouted, and Agnethe and Karen ran over to me, as the smashed bottle tumbled down into the cellar. They helped me up the stairs, and we sat by the kitchen window, surveying the damage. I screamed again when we noticed that a large piece of glass was still sticking out of my finger. Agnethe ran to fetch Benze, who came hurrying along. She took me over to the kitchen sink and told me to hold the injured finger, which was dripping blood, over the sink, whilst she found a pair of tweezers. Benze removed the piece of glass and rinsed my finger, looking to see if there were any more glass shards in it. Then she bandaged my finger tightly, so that I wouldn't have to have it stitched back together. I had to hold my finger for a while, and it throbbed. There were drops of blood all the way from the basement stairs to the utility room. Karen rushed down to the basement, picked up the zinc mop bucket, put in brown soap, and washed the floors in the kitchen and on the stairs. I could only watch for the rest of the baking session.

During the month of December, we baked cakes, brownies, shortbread, and cookies. We had to stand back when Karen put the baking trays into the oven; it looked red-hot, and Karen's face would flush with the heat that emerged from it. Once we had done our homework in the afternoon, we had a relaxing lesson where we learned Christmas hymns, did Christmas drawings, and enjoyed a cookie whilst the blizzard raged outdoors. We felt safe and cosy inside. On a couple of evenings, we had rice porridge, with cinnamon sugar and buttercream. One time, some of the bigger boys got some dark rye bread from Echwald the baker, and Karen cut it into thick slices, using the small hand bread cutter. We heard the sounds of the slicing coming from the pantry – it was a tough job, which we weren't allowed to help with, as it was too dangerous. The bigger boys helped Jens to shovel the snow, and the rest of us worked on changing the bed linen, and doing the rest of the big pre-Christmas clean-up. Jens was also busy, making our first Christmas tree stand.

Christmas and New Year

Christmas packages from our Danish foster parents had arrived on the last ship of the year from Denmark, and we were sent out to play when we got too curious about them. Unless they contained fresh fruit – which Benze could smell – the packages had to go up to the toy depot. I got apples from my foster parents at the Brøderup primary school, but unfortunately, many had rotted in transit. Curiously, we watched what was going on, through the garden door on the terrace. First, Miss Blom helped Benze to unpack the outer brown wrapping paper; once that had been done, they placed the individual packages, which were wrapped in Christmas paper, into a big sack. We moved around a bit, to recover the warmth in our toes. Suddenly, four stools were put out on the terrace, and a large cardboard box, filled with half-rotten apples, was placed on top. 'Can we have some?' we asked, and yes, we could eat as many as we wanted to. In mere moments, next to nothing was left – only the most rotten apples of all. The older boys even ate the apple cores, even though some people said that apple tree branches would grow out of their ears. I tried to eat the cores, too; I thought that the seeds tasted nice.

Just before Christmas, a little boy came running in. He looked neglected. He was sniffling and seemed to be frozen; and when he saw us, he ran in crying to Karen, in the kitchen. Karen sent him out again, placing a kiss on his lips with her index finger. But he went back in again and clung onto her skirts. We felt warm all over when Karen spoke to him in Greenlandic: '*Aataakkunnukarniarit*'

('Go over to your grandparents'). We talked about the visit from the boy stranger. Miss Ingrid had heard the disruption, and called for Benze, who rose sleepily from her post-dinner nap. We stood watching what happened. When Benze came in, the boy was still crying and wouldn't let go of Karen. We thought it was disgusting when he wiped his nose on her skirt. He cried, heartbreakingly, still holding tight onto Karen's legs. We were sent out of the kitchen. The next day, the boy moved into the orphanage. It turned out that it was Karen's five-year-old son, Helge. Benze hadn't known anything about him previously, and she felt sorry for him. Helge was allowed to move in to a room with Little Karl, rather than being allowed to sleep in his mother's room. Benze said that there wasn't enough space, but we sometimes saw Karen taking Helge back to Little Karl's room, after he had fallen asleep in hers. Helge was always at his mother's heels; he cried a lot, and he couldn't understand much, so we teased him, as we teased Karen, who didn't speak much Danish. One day it became too much for her, and she chased us around the tables in the dining room, waving a towel around in the air, and scolding us in Greenlandic. Giggling, we ran out into the hallway, and back to our rooms, where we held the doors shut tight. Benze heard the noise, and Karen disappeared into the kitchen. When Benze looked in to our rooms, we were already sitting there politely, looking at our albums with the glossy pictures.

One evening after dinner, there was a nice surprise – we were going to make Christmas decorations. A delivery of shiny paper, cardboard, scissors, glue, model pixies, Santa Claus models, and pastries had arrived on the ship. I found it easy to make the paperchains, and to cut out the angel and pixie shapes, but harder to learn how to make the heart-shaped Christmas decorations.[17] We put what we'd made into a big box, ready for Christmas Eve,[18] when the adults would decorate the Christmas tree. The snow fell every day; on some days, there were more snowflakes than there were on others. We threw ourselves back into the snow and made our snow angels. It was wonderful just to lie there, looking up at the falling snowflakes, enjoying the Greenlandic winter, feeling in every fibre of my being that I was home again – how lovely it was! But when the wind started to blow, there could be some serious snowstorms. The electric lights would go out, and we'd use the kerosene lamps. If we ran out of oil, Benze would order a bulldozer from the GTO (Greenland Technical Organisation), so that the tanker and electrician could reach us. On days like these, confusion reigned.

Sometimes, after a blizzard, snow would pile up in big drifts outside the orphanage. We would go out with shovels and build caves in the snowdrifts. Later on, we were allowed to put candles into the smaller of the hollows we'd made in the snowdrifts, and we would stand inside, behind the windows, and watch the lights flickering there. That was nice. We were allowed to take a carrot, for a snowman's nose, two pieces of charcoal for his eyes, and smaller pieces of charcoal to make his mouth and eyebrows. In the winter, we usually used the basement door to get in and out. When we came inside, we could hang our wet clothes on the clotheslines in the basement and stand our boots in a row in front of the basement fire.

We had to change into nice clothes for Christmas dinner. Before beginning the meal, '*The Little Matchstick Girl*' was read to us. Nobody at all was allowed to go into the dining room until everything was ready; even the adults. If one of them had left something in the kitchen, instead of going through the dining room, they had to go up the stairs, all along the hallway on the first floor and then down into the kitchen that way. Everything was top secret. But finally, the dining room door was opened: 'Welcome!' The Christmas tree had electric lights, and there were candles on the tables with the white damask cloths, a soda at each seat, wine glasses for the adults, and an almond present[19] at every table. We couldn't even fit all of our gifts underneath the Christmas tree. There was a smell of roasted meat, and we looked at each other, and at the adults, who instead of wearing uniforms, were wearing nice dresses, and had big smiles. It was a great Christmas dinner. There were rice puddings,[20] with red syrup; I don't remember who won the almond present, which was a marzipan pig, with a red bow on his stomach. After Christmas dinner, we could go back to our rooms, or into the hallways to play, whilst preparations were being made for dancing around the Christmas tree. We could hear the blizzard raging outside. Then the bell rang, and the dancing started. We started with 'A Child is born in Bethlehem', and continued with other Christmas hymns, and finished up with 'From the Top of the Green Tree'. Then we were told to find ourselves a place to sit down to unwrap our Christmas presents.

We were very happy on our first Christmas at the orphanage. We got so many Christmas presents from Denmark. Søren got a big fire truck, and Barselaj shouted out, 'I'm going to be a fireman when I grow up!' Aaron followed this up by shouting back, 'Well, I'm going to be a cop!' I was standing just to the right of the French doors, with my gifts on the small table by the radiator. My foster parents had sent me a big bag of sweets. 'Can I see?' said Benze, and then went into a long explanation of why she should probably take care of them, because I shouldn't eat them all at once. 'You can have one now, and then I'll give you one each day'. I felt really cheated – she'd just taken my present! I felt like sobbing; my throat was tight, but I pulled myself together, because I didn't want to cry. 'It's Christmas Eve, after all', I thought. As I was unwrapping my next present, we could hear the blizzard howling even louder than before, and suddenly, the light went out. We screamed, the youngest amongst us cried, but I stayed rooted to the spot, holding my arms tightly around my presents. Benze fetched a kerosene lamp and collected the candles from the kitchen, so that we could see our Christmas presents again. When the adults had coffee, we had more soda and then nuts, Christmas sweets and oranges were put out onto the tables. The last thing I thought about before I went to sleep was how my mother and siblings were spending Christmas. They were so close, but I didn't know that then.

On Christmas Day, we were going to see the Christmas tree in the big, brown American Consulate building. Benze went in first. We girls wore princess dresses, and the boys wore their white anoraks. When we went inside, we had to take our coats off and put on indoor shoes. First, we went through the dining room, which was a few steps up from the big, beautiful living room. Then we saw the biggest Christmas tree that we had ever seen. The living room had a very high ceiling, and the Christmas tree looked twice as tall as ours. We danced around the tree, and afterwards, there was a big bag of goodies for each of us. While the adults were having coffee, we watched '*Far til fire*'[21] in the Consulate's own cinema.

The next Christmas Day, we went to see police chief Simony and his wife's Christmas tree, in their large residence at the station in Old Godthåb, with its smell of Christmas baking. My mother's friend, Birgitte, worked as a *kiffak* at the Simonys'. She patted me on the head and whispered, '*Iliinannguaq*' ('sweet Helene') in my ear. We could play all over the house. Next to the entrance to the kitchen, there was a staircase up to the first floor, and we all ran up and down stairs, over and over again. There was a view out over the sea from the living room window. It was too stormy for us to go out to play. We danced around the Christmas tree and got goody bags here, too.

On our third Christmas Day at the orphanage, we went down to the Governor's residence. This was the tall grey building in the colony harbour, next to the church. Benze was very popular there, and all of 'her children' were made welcome. We saw the Christmas tree there, too, and afterwards we each got a bag of sweets and an apple. We girls took our time over our goody bags, so that we could enjoy the last few bits of it at home in the evening. It was fun to make the boys jealous, because they had already eaten theirs.

On our fourth Christmas Day, those of us who had mothers in the town were allowed to visit their families, in the afternoon between two and four o'clock. It was strange to be permitted to visit my mother. I stepped through the door, into my childhood home. My mother, Victoria, and Hans were smiling warmly. My Mom took me kindly by the hand, showed me into the living room, and pointed at the couch. I sat down, and Victoria and Hans came and sat down next to me. The house was decorated for Christmas, with garlands hanging from the ceiling, and a small red spruce Christmas tree, with candles and paper hearts. Hans went over by the Christmas tree and opened the basement door, to show me that they had got raisins and prunes. Mom ushered him back to his place and said, '*Uatsilaaq*' ('Wait a minute'). Mom had baked a Greenlandic cake, with lots of raisins, and we had tea with milk in it. 'Have some', said Victoria, pointing at the cake. We took a slice each, putting them on side plates. There were also sweets and sugary treats that had been fried in a pan. I remembered these from before I had been taken away. We enjoyed this treat in silence, smiling at each other. Once in a while, my Mom would say, '*Iliina*',[22] smiling shyly at me. We also danced round the Christmas tree, and I recognised '*A Child was born in Bethlehem*' from its tune. But when we were all cuddled up, my mother pointed to her alarm clock, and said, '*Gloggen fia*'. She meant, 'Four o' clock',[23] and time for me to go back to the orphanage. I ran home, and that night I cried myself to sleep. I wondered why, if there was room for me at home, I had to stay at the orphanage.

After lunch on New Year's Eve after lunch, we were setting the tables for the evening. We spread the freshly ironed white damask cloths out on the tables and hung up streamers from the light fittings. Also, we placed a couple of sparklers next to each plate, party poppers on the cake plates, and small New Year's hats all around. We were having cod for New Year's dinner. 'Oh, why do we have to eat cod at New Year?' we asked each other. We didn't eat that much of it; but we did eat a lot of dessert, which was chocolate pudding. After dinner, the tables were pushed to the sides of the room, and Ane Sofie and Little Karl's gramophone was pulled up, so that we could dance to the gramophone records that they had been given by their foster parents in Denmark. Some people didn't want to dance, but we all played together with our new toys. We stayed up very late. At eight o'clock in the evening, we went out and lit our sparklers. They burnt brightly, and afterwards we were allowed to put candle-ends into the little hollows that we had dug into the snowdrifts. As they were sheltered from the wind, the candles burned for quite a while.

'Did you hear that?' We looked at each other and listened hard. 'Look!' we cried. It was shortly after New Year 1953, and suddenly, we saw an aeroplane in the sky. 'It's from the US base', said one of the adults, who had come out when they heard the alarm. From the orphanage, we had a view of the marsh. We ran down the end of Børnehjemsvej. The American aeroplane had dropped several boxes into the marsh. It looked nice; it was the mail for the American Consulate, which had been dropped via several different coloured parachutes. 'It's good that there's a lot of snow on the ground, so the mail wasn't get damaged', I thought. I remembered then that I had seen an aeroplane like this before, when my father was still alive.

A Hello from Godthaab

One day, a very large, elongated package was delivered to the orphanage. We were right on Benze's heels when she unpacked it in the kitchen. 'There it is', she said. She had been looking forward to the delivery of what turned out to be a replacement for the rye bread slicer. Karen now had difficulty in using the old slicer to cut our rye bread every day, and now this semi automatic bread cutter had arrived. We were told not to touch it. After cleaning up the packaging, Benze and Miss Blom went into Benze's apartment, to have 'a couple of drops' of coffee. Karen went up to her room with a cup in her hand and also had a coffee break. The door to the pantry was open. Agnethe and I went in, to study the strange contraption more closely. I went up to it rather cautiously, but curiosity soon overtook my initial apprehensiveness. I saw the rye bread on a triangular shelf. I picked up a big, freshly baked loaf, and tried to work out how the sharp semicircular blade turned. There was a wooden handle which was sticking out from the right-hand side of the blade, but I couldn't reach it easily. I heard Agnethe say, 'We musn't!', but in my excitement, I ignored her warning. With both hands, I tugged at the handle; the blade was sharp, and heavy, and suddenly, the big handle came down. I felt a sting and Agnethe gasped as I pulled my hand back. To our horror, I could now see the little finger bone of my left hand and my torn flesh hanging loose. I could feel the warm blood seeping down to my wrist. Using my right hand, I quickly pressed the flesh of my little finger back together. Agnethe ran to fetch an adult, shouting, 'Helene has cut herself! Helene has cut herself!' Now pale, I laid there, my back pressed against the window in the pantry. Benze, along with a whole crowd of the children, ran in. 'What happened?' asked Benze. 'Nothing', I whispered. Then Benze looked at the blood on the machine, and the drops on the floor, and then back at me. By now I could no longer hide the hot blood trickling down my left arm. Benze gently pulled my hands apart and then quickly pushed them back together again. No more questions were asked; I was rushed out into the long corridor, in front of the medicine cabinet. Benze quickly opened the white doors of the cabinet, and without saying another word, she dressed the wound tightly, holding my flesh in place. She then told me to raise my hand, and she bandaged it all the way up to the wrist. I was completely silent throughout this; I didn't cry at all. Fortunately, I did not have to go to hospital, and afterwards, the other children all admired me. But my little finger hurt a lot.

Some of us girls were following our new *kiffak*, Birthe, as she went around the rooms, changing the bed linen. 'How do you say "bed" in Greenlandic?' we asked. '*Sinniffik*', she replied. That sounded easy enough, we thought, and we giggled and repeated the word. On the way up to the big rooms, we asked her, 'How do you say "stairs"?' '*Majuartarfiit*'. We laughed, whilst trying to pronounce it. Benze came straight out of her apartment, followed us up the stairs, and asked us what we were doing. 'Are you trying to learn Greenlandic?' she asked us, sharply. 'Yes, from Birthe', we giggled. Suddenly, we noticed Benze's sharp expression. 'You can stop that immediately', she said. Then, turning to Birthe, she said, 'I never want to hear you trying to teach these children Greenlandic again. They are Danish-speaking, and they must continue to be so. Is that understood?' We got sent out to play, and the Greenlandic *kiffaks* were called in to a meeting in the dining room, and Benze spoke in a strange way to them. A bulge in her neck seemed to be forming. I was deeply shaken, and my thoughts ran free. 'But why? Then I won't be able to talk to my mother and my siblings again'. I felt completely empty inside. Some of the others came, and asked me what was wrong. A small group of us sat down on a couple of toboggans, looking absolutely abandoned. I was really angry at Benze. 'That stupid old woman! She seriously thinks she's our mother, but she's more like an old granny!' At dinner, Benze tried to make eye contact with me, but I ignored her. I nodded gently at Birthe, but she hardly dared to look at me.

Benze got a new radio. It was like a small closet, with speakers surrounded by a thin, decorative wooden frame. The programme buttons were at the base, and a light came on when the radio was switched on, so that you could see if the dial was turning to the right or to the left. A long lead ran from the rear of the set into the dining room. A speaker was placed just above the cupboard in the serving wing of the dining room, so we could hear the radio there, too. It annoyed us when Benze wanted to listen to the radio news, because she turned volume right up, and we all had to be quiet. On the slightest disturbance from us, she'd lift her index finger to her lips, with a loud, 'Shush!'

Most of our *kiffaks* spoke Danish very poorly. They were not alone in that. You could hear that when I listened to the requests programme on Sundays. One Sunday, as I was tidying my little cupboard, greetings from near and far were being sent out on the radio. I strained to listen as the female broadcaster seemed to read out a message which ended, 'Loving greetings from Indochina'. 'Hold on a minute', I wondered, 'Who on Earth knows anyone in Indochina?' Then she read out another message, again ending with 'Loving greetings from Indochina'. Suddenly I burst out laughing, and the other people in the dining room asked me what I was laughing about. When I'd caught my breath, I told them about the messages, and what I'd misheard: 'Loving greetings from someone you know'.[24]

Benze had to keep track of many things. When we moved in, the electricians had not quite finished the installations. Some of them were setting up an electricity pole outside, next to the basement door. We were all watching – the boys were outside, and some of us girls had opened the window and watched from inside. We thought it was funny when one of the electricians put what looked like crab claws on his shoes, so that he could climb up to the top of the post, and screw in the white porcelain heads, on which the overhead lines would be fitted. 'When the last of the poles have gone up, then we'll need to power up the mains', one of the men said. The tradesmen enjoyed talking to us, because we spoke fluent Danish. An electrician climbed up the pole and put his hand on the wire. When he touched it, he jerked, and suddenly, he was lying on the ground. We were completely silent. One of the *kiffaks* went to fetch Benze, who ran to the scene as quickly as she could. She asked one of the *kiffaks* to run and fetch the doctor, and the boys moved aside. We girls hurriedly closed the windows. Later on, we heard that the electrician was dead. It made a deep impression on us.

The medical room was used a good deal. I don't remember how often the doctor came and examined us, but when he was there, we had to stand in line in front of the medical room in our underwear, and then we were brought in, one at a time. We had to be measured and weighed; the doctor listened to our chests; we had to open our mouths as wide as we could, so that he could examine our throats; our teeth were checked, and we had to cough. Miss Blom stood, and Benze and the doctor sat down. 'We are not even sick. Why do we be examined so thoroughly?' I wondered.

Just after the telephone was installed, journalists from the Red Cross and *Billed-bladet*[25] visited the orphanage. I think that they were the same people who photographed us whilst we were in Denmark. Someone took a photograph of Barselaj, standing at the small telephone shelf in front of Benze's door, holding the telephone next to his ear, whilst saying, 'Hello'. The text beneath the image read (Figure 15):

> *A hello from Godthaab.* This picture is a scene from Greenland, where there are also state-of-the-art telephones now, of the same kind that are gaining ground in Denmark, alongside the current full automation process. The

picture is from the new Red Cross orphanage in Godthaab. One of the small residents, a Greenlandic boy named Barselaj, is trying to understand the mysteries of the telephone. The city of Godthaab, the capital of Greenland and the site of the Governor's residence, now has a fully automated telephone system, thus replacing the old magnetic switchboard model. Whilst telephones are not needed at these latitudes to the same extent as they are in Denmark, Godthaab's half-hundred subscribers include administrators, tradesmen, the school system, the health care system and the technical service, who sometimes have some things to talk about. There are no private telephone quarrels! The fully automated system, which is now being implemented in Godthaab, will probably not be complete until next year, as it is often difficult to lay cables and fix lines, due to the rocky terrain.

There was an unmistakable stench when 'Crappy Jens' emptied the toilet bowls. They were always so full that it seemed that he was constantly running up and down the stairs. We were both quick and enthusiastic to run out and play on those days. Fortunately, this situation didn't last long before the plumbers came. We would be amongst the first people in Godthåb to have flushing toilets. There was a terrible noise with all the craftsmen, but how nice it was afterwards, with clean, almost odourless toilets; there was even room for a small sink next to each toilet. But it was annoying when you couldn't pee in peace. The boys had developed a habit of listening out for us girls, then lying down on the floor, and looking up at us from underneath the locked door. Suddenly, we would hear them giggling and see their naughty faces. We would scream, quickly pull our pants up, and then run out and gossip with each other about the situation. Warnings were given several times before meals, and when these things were mentioned, we girls would stare hard at the boys.

Bodil always fell asleep very quickly, and sometimes it seemed that she sleepwalked. It scared me and Ane Sofie a bit. 'Does she really walk in her sleep?' we wondered. We wanted to see this for ourselves, so Ane Sofie agreed that we'd get up in the night, after Bodil had fallen sleep. I felt confused when Ane Sofie woke me up, holding a finger to her lips, but I rubbed my eyes, and listened. I did the right thing – Bodil was in her flannel nightgown, on her way to the toilet, with her arms stretched out in front of her. She sat down too far down into the toilet bowl; we giggled, and Ane Sofie gently pulled her back into position. Then she peed. Without drying herself, she pulled up her pants and went back to sleep in her bed. Ane Sofie and I looked at each other, giggled, and went back to sleep. The next morning we told her she walked in her sleep, but she denied it. We kept on at her, though, so eventually she got mad at us and reported us to Benze. Benze told us not to tease Bodil, and we were both angry about this.

'Bath time!' came the call. 'Oh no', we thought, 'A communal bath again'. Of course, Benze wanted to make savings here, too. The bathtub was filled, and we had to change all of our clothing, putting the washing out in small piles in the hallway. This was no fun at all. Eventually, we didn't even have many of our

ET HALLO FRA GODTHAAB

NÆH, billedet er skam taget i Grønland, hvortil der nu ogsaa er kommet hypermoderne telefoner af den type, som er ved at vinde indpas i Danmark i forbindelse med helautomatiseringen.

Billedet er fra Røde Kors' nye børnehjem i Godthaab. En af de smaa beboere, grønlænderdrengen Barselei, prøver at sætte sig ind i telefoneringens mysterier. Byen Godthaab, Grønlands hovedstad og landshøvdingens residensby, faar nu helautomatisk telefonsystem til afløsning for det gamle med magnetocentral. Der er ikke paa de breddegrader brug for telefon i samme udstrækning som i Danmark. Godthaabs halve hundrede abonnenter er da ogsaa administrationen, handelen, skolevæsenet, sundhedsvæsenet og den tekniske tjeneste, som jo af og til har noget at snakke sammen om. Privat telefonkævl er der ikke noget af!

Det helautomatiske system, som nu gennemføres i Godthaab, vil antagelig først være helt færdigt til næste aar. Der skal lægges kabler og trækkes luftledninger de steder, hvor man paa grund af klipperne ikke kan gaa i jorden.

FIGURE 15   A 'hello' from Godthaab

own clothes – only our dresses were our own. Why do we have to bathe with the boys? We stood in a row and started by washing our hair in brown soap in the small washbasins in the bathroom. A *kiffak* helped us. Afterwards, we had to soap ourselves all over, and our hair was rinsed in vinegar water. Finally, we had to go down to the bathtub and rinse ourselves off, and then dry ourselves. There was a lot of noise and pushing, giggling, and laughing. Afterwards, we had to go to the closet and try to find some clean clothes of our own sizes.

New Children

There was a message at dinner. Eli and Ole had to move into the two-person room next to the ironing room, because three new children from Holsteinsborg were to join us. 'They're going to move in, so we'll have to move some of you around', we were told. We children became quiet and looked at one another. 'Why?' asked Ole. 'They are three siblings who need a new home. They've just lost their parents', came the reply. 'What are their names?' we asked. Pressing her palms on the edge of the table in from of her, Benze replied,

Julie, Dorthe and Karl-Otto. They cannot speak Danish, but we must teach them how to do so, quickly. Miss Else Petersen, your dance teacher, will come here to the orphanage on Mondays, and start giving them additional lessons in Danish. Those of you who are not so good at writing in Danish should also join these extra classes.

What a mouthful. As soon as the meal was finished, Eli and Ole started to move their things. They were grinning all over their faces at the prospect of having the two-person room.

Then one day, during dinner, Julie, Dorthe, and Karl-Otto came in, through the French doors. Julie was ten years old; Dorthe was eight years old; and Karl-Otto was five years old. We all peeked at them; they looked like they were scared to even breathe. They only had one suitcase between them. Finally, Miss Ingrid went over to them, and welcomed them, in Greenlandic. A palpable hush descended. They went into the living room, and we watched every step that they took – they still had their rubber boots on, and they made wet spots on the floor. I felt deeply for them. I thought to myself, 'What a shame it is for them. They have lost both their parents, and now they have to live in an orphanage'. Inwardly, I relived the time that I was sent away from my mother and remembered how sad I had been. Dorthe was allocated to our room. She was the same age as Agnethe and me, and we took her under our wings. That first night, she was cried herself to sleep, and her feelings were infectious. We soon got to work on teaching our new 'siblings' Danish. Sometimes we laughed at their pronunciation, but if anyone else laughed at them – say, our Danish classmates – we would angrily protect them. When Benze's friend, Miss Else Petersen, started giving the extra Danish classes, we were told that everyone should take it in turns to join the class, alongside the new children. Miss Else Petersen had to assess whether there were others amongst us who needed this extra tuition. It was so boring! In every lesson, she made a list of words that we had to say out loud, and memorise. 'Whoever spells "done" the wrong way, will get his backside slapped today'. Yawn! It was just like being at school. Luckily, I only had to go on one occasion.

It was much more fun, of course, to spend my time playing with my dolls or swapping glossy pictures with the others. That was one of the things I spent my money on when we started getting pocket money every week. The children in the first grade got fifty øre, those in the second grade got one krone, the third graders got one krone and fifty øre, the fourth graders got two kroner, and the fifth graders got two kroner and fifty øre. We each got a box and a savings book, and we had to put half of our pocket money into savings. It was hard to this. But it was very exciting to go into the little green building of Ole's Department Store. At the beginning, there were not too many things on sale, but later on, they had lots of glossy pictures, and doll's clothes. The dress-up dolls were quite expensive, so I had to save up a long time in order to buy them. I already had one dress-up doll, which was a gift from my foster parents in Denmark, and then I had my china doll, Tove, who was my greatest comfort – she helped a little, when I was missing my foster family. Bodil, Ane Sofie and some of the other girls also had beautiful dolls, and their heads, arms, and legs were made of porcelain. The baby hair was painted on my doll's head, just like the eyebrows and red cheeks, but the eyes could open and close. We often played 'Mommy, Daddy and babies' with the dolls. One of our joint Christmas gifts had been a toy oven, with pots and pans, which we used in games like this, which we usually played in our rooms. Occasionally, the boys would come barging in, and interrupt our games. If Søren and Gâba were amongst them, we'd shout at them and call them names: 'Get out, Chinaman and Elephant Ears!' Of course, we got the same treatment back from them: 'Okay, Big Teeth and Black Butt!'. I hadn't noticed before that my new front teeth were so big, but now I was embarrassed by them.

It wasn't too long afterwards that we were told that two more children were coming. It was a brother and a sister – Regine, who was nine years old, and Samuel, who was six years old. They came from Fiskenæsset, which was a place near Godthåb. They didn't know Danish, either, and they had a lot of difficulty in learning it; they fell well behind the three children from Holsteinsborg. After Regine and Samuel joined us, there was no room for anyone else – all of the rooms had been filled. Extra racks of coathooks were put up in the wardrobes, and two extra chairs were ordered. Gabriel and Samuel were difficult names for us to say, so pretty soon they became known as Gâba and Sâmo, instead.

We had learnt to sing, 'Shrovetide's my name, and I want cakes!', and we were all busy preparing the cakes, cat sweets, and decorations.[26] Benze had already ordered the Shrovetide rods in advance, and they were hanging from the ceiling. Benze had lived in Greenland for many years, and she was well used to planning ahead – often, a year in advance. So what wasn't in the various stores around the orphanage was probably not worth mentioning! We had made a secret plan with Miss Blom – we would creep into Benze's room in our nightwear and wake her up in the morning. We went along the hallway and quickly crossed through Benze's living room, and then into her bedroom, where we pulled off her bed-clothes, screaming, 'Shrovetide's my name, and I want cakes!'. She looked startled, and we thought that this was very funny. In the afternoon we dressed up, painted our lips and our cheeks red, painted on black eyebrows, and put on our little hats. Then we played 'hit the cat out of the barrel' in the dining room. We grabbed at the little bags of sweets and treats, as they came shooting out of the barrel. Afterwards, we had freshly baked rolls and hot cocoa.

Miss Ingrid

Miss Ingrid Holm was seventeen years old, and her father was the supplies manager in Kapisillit. She was the youngest of the *kiffaks,* the most beautiful, and the one who could speak the best Danish. At that time, Kapisillit was known for scallops and cod fishery (it was called 'Greenland's Lofoten'),[27] and at the same time, salmon spawned in the river near Kapisillit. Miss Ingrid and our Danish child nurse, Miss Birthe Blom, were the first people other than Benze to be employed at the orphanage. Miss Ingrid was paid 125 kroner per month, and whilst Miss Blom got twice as much, they both did the same amount of work. Miss Blom had previously worked in Police Chief Vesterbirk's house in Godthåb. Later on, she married a commercial manager named Egebjerg. Miss Ingrid was with us all of the time, but Miss Blom spent her evenings at her Danish friends' homes.

Benze was a tough employer. All of the furniture, all of the rooms, and all of the bedding had to be washed prior to our arrival on September 25, 1952. When the orphanage was being built, Miss Ingrid and Miss Blom were set to work, cleaning the cracks between the floorboards, one by one, with a pair of old broken scissors. Afterwards, the floors had to be scoured and washed. For that task, they had been given two buckets – one with hot water and brown soap, and the other with plain cold water. Finally, the floors had to be painted. Benze had previously been a head nurse, and Ingrid thought her nickname was very appropriate – behind her back, people called '*Oorussuaq*', which, freely translated, means, 'the one who always whips up a troubled mood'. The same type of attention to cleanliness applied to us, when we were bathing. First of all, our hair was washed with brown soap and, afterwards, rinsed with vinegar water. Benze explained that this was so that we would not get lice. Our hair was always shiny. At the beginning, it was Miss Ingrid who cut our hair, after our bath. The orphanage was quite messy when we arrived. The tradesmen had only just managed to get the washing machine installed.

Miss Ingrid was good at Danish because in 1947, as an eleven-year-old, she had been sent to Denmark. She had been really sad to have to travel so far away and had cried when, on her departure, her mother had said to her, 'We'll see each other in three years' time. We're doing this for your own good'. When Miss Ingrid got to Denmark, she was sent to the Frankrigsgades School, where they had to sing hymns every morning. She felt very honoured when Queen Ingrid, who was a patron of the school, came to visit. The next time that Miss Ingrid went to Denmark, she attended the Jægerspris School, and she was the only Greenlander there. During the time that Miss Ingrid was at school, she also took dancing lessons, and learned the tango, quickstep, English waltz and the foxtrot. She also earned a certificate in lifesaving when she was in Denmark.

In 1952, Ingrid returned to Kapisillit, which was the same year that the first reindeer were brought to Greenland. They were transported on a ship called the *Martin S*, which docked at an old sheep breeding station. The new reindeer station was therefore in Itinnera, which is located directly opposite to Kapisillit. When there were larger ships in Kapisillit during the summer, Miss Ingrid loved to jump about in the waves. The crew aboard the ships admired the beautiful young girl, with her big blue eyes, freckles all over her face, and her charcoal-coloured, wavy hair. They all stood against the rails of the ship clapping and whistling at her. Miss Ingrid's father, however, was angry over her swimming displays. The rumours about her had reached the National Council, who called her 'the swimming girl'. Miss Ingrid's mother was the daughter of Anne Marie Holm Chemnitz, who was a nurse. Her father had invented '*kussartitat*'. This was a method of wind-drying fish, which was accomplished by tying the fish together in pairs and hanging them out to dry on tall racks. Later, the KGH took up this idea, and it became widely used method throughout Greenland. 'Little Per', or Per Berthelsen, recorded a song which told about the dried fish being sent to Japan. Miss Ingrid had several family members in Godthåb. Amongst the people there, her family was known as '*angisoorsuakkormiut*', which means, 'those of the big people', although Ingrid herself was 170 centimetres tall. Her paternal uncle had a sheep farm in Godthåb. When the sheep bore lambs, they wandered about grazing, all around Godthåb. The ram was tethered by the water's edge at the Holm's place. When there were too many sheep – they had begun to knock the garbage cans over – a ban was placed on keeping sheep in Godthåb.

Miss Ingrid returned to Greenland as a seventeen-year-old, in order to learn Greenlandic again. She had wanted to be a hairdresser, but when she was stalled in her ambition, she gave it up. So instead, she started work as an interpreter for the specialists within the GTO, and other organisations. In the summer of 1952, the royal couple, Queen Ingrid and King Frederik, came to visit Godthåb. Ingrid Holm, together with Augo Lynge's daughter Astrid Najaaraq, helped in the presentation of Greenlandic national costumes to the three little princesses, and Queen Ingrid. King Frederik received the national costume for Greenland men – a white anorak. During the royal visit, there was dancing, music, and partying for everyone, in the old Town Hall. There were lots of people there, and when the music started up with a Greenlandic polka, there was a huge banging noise, and the floor collapsed. Fortunately, no one was hurt, and later on, a new floor was laid. That was the point when the first carpentry stores opened up and spread throughout Greenland.

One day in 1953, when Miss Ingrid was at dinner with Principal Binzer, she met two tall Danish brothers who had come to Godthåb, in order to see and experience Greenland, and she fell in love with them. They were Hans and Per Buckhøj Sørensen. One was a house painter, and the other was an artist, and they had taken up temporary work as garbage men. At the dancing after dinner, they had taken turns dancing with Miss Ingrid. They were surprised that she was so good at Danish and that she was such an amazing dancer. She had then told them about her time in Denmark. The brothers then got the idea of starting dancing lessons in their spare time, and they asked Miss Ingrid to join them. She agreed, and they met up a few days later, in order to make the announcements to the town that they would be teaching the various different styles of dance, in the assembly hall in the evenings. There was great interest in this. Amongst the first people to sign up were some of the students from the college. Hence, many of Miss Ingrid's hitherto free evenings were quickly filled, with dancing lessons and performances in the tango, quickstep, English waltz and the foxtrot. Miss Ingrid was not paid for her teaching, but in return for it, she had her dance shoes re-soled for free four times. At the orphanage, we also enjoyed the benefits of the dancing classes. Miss Ingrid persuaded Benze to allow the Buckhøj Sørensen brothers to visit and to teach us how to dance.

We had met already met Queen Ingrid, of course, when we were in Denmark, as she was a patron of the Save the Children. After we were moved to the orphanage in Godthåb, it seemed that she was still following us. One day, a big package arrived, which had been sent by the Queen. It contained princess dresses, for each of the girls. They were exactly like the ones that the three little princesses – Margrethe, Benedikte, and Anne-Marie – were wearing in one of the pictures of the royal family that we had seen. They were tiny, pink dresses with a small collar, and a wide skirt, which fastened at the back. For the boys, there were burgundy-coloured checkered anoraks, plus-fours and knee socks. We wore the princess dresses on the Sundays that we had the dance lessons, and when we were having our photographs taken. We had dancing lessons for several consecutive Sundays when we were preparing for a show at the Assembly House. Sometimes, important guests were allowed to come in and see us on the afternoons when we had dancing lessons in the dining room.

The Danish School

Jørgen Fleischer wrote in his book, '*Forvandlingens år: Grønland fra koloni to landsdel*' ['The Year of Transformation: Greenland from Colony to County']:

> On March 28th, 1950, the Greenland Commission's report was published in the Danish press, and discussed extensively in a number of important articles. The reorganisation of 1950 meant a number of radical changes in the administration of Greenland. Greenland became a department under the Ministry of State. The Administration and the KGH, were now no longer under the same roof. KGH was transformed into an independent trading company, but retained universal service. The state monopoly was thus abolished, and Greenland was no longer a closed country. There was a separation of schooling from the church, and teaching was expanded, with specific Danish-language primary schools in the colonies, where the students could sit examinations similar to the Danish versions.

All of the children at the orphanage were enrolled in the Danish school in Godthåb, but being as there were more than enough students to fill the classes, the mixed (Greenlandic/Danish) married couples with children were asked if they were interested in enrolling their children at the Danish school. There was no real difficulty in persuading such parents, but because the class quotas were small, it was sometimes the case that children from two different school grades had to be taught in the same classroom. This is why I was in the same class as Vivi, whose mother was a Greenlander, and sat on the Social Committee at that time. Both Vivi and her big brother, Ejvind, often came up to play with us at the orphanage. Ejvind was in the same class as Eli, who was the oldest of our boys. Their mother, Guldborg Chemnitz, was one of the guests who accompanied us in our activities at the orphanage, and Vivi and Ejvind's grandmother was also a significant woman in Greenland's history. Katrine Chemnitz started the Women's Association in Godthåb and was the first female member of the Social

Committee. That's why she was one of those people who often watched when we had our dance lessons.

Miss Ingrid told us that both the Greenlandic and the Danish residents of Godthåb called us 'those spoiled orphanage children'. But this wasn't a bad thing. When we were out walking, usually two-by-two in a line, with our neat hairstyles and clothes, speaking Danish, looking happy, and singing Danish songs, both the local children and the adults seemed to envy us. We did well at school, and the local children were ashamed in front of us, because we could speak Danish fluently, ate good food, and had nice clothes. One time, Miss Ingrid tried to spread margarine on our bread, and was scolded by Benze, who told her that we were to have Lurpak butter on our rye bread instead. In fact, it was an enviable thing to be at an orphanage in Godthåb in the 1950s.

Our daily schedule was fixed: the morning bathroom routine, then breakfast, then school, an activity after school, going out to play, coming inside to do one's homework, dinner, being read aloud to, out playing again until about 8:00 p.m., the evening bathroom routine, kneeling down in front of our beds to say the Lord's Prayer, and finally, 'Good night'. When the weather was bad, and we couldn't go out to play, we had to play indoor games, or football in the attic room. Miss Ingrid was always at home, so sometimes, we'd visit her room, and listen to music, and dance.

One time, Benze suddenly announced to me that Miss Ingrid had an hour to spare, so she could take me to visit my mother and my siblings. I felt both happy and distressed, but most of all surprised. So, now it was time to go home for a short visit. 'Does my mother know something?' I thought, as Miss Ingrid and I walked hand-in-hand down Børnehjemsvej. 'Can she understand more Danish now? And what about my big sister? My little brother probably can't understand Danish'. From Skibshavnsvej, we crossed behind Uvdloriánguak Kristiansen's house, and there was my childhood home. I saw that the lights were on, so my family was at home. Miss Ingrid knocked on the door, and my mother opened it, smiling and surprised. '*Tamassa iserniaritsi*' ('please come inside'), she said. Victoria and Hans came running in from the living room and said a great deal in Greenlandic. We smiled at each other. I felt alien, but at the same time, touched by finally coming back to my childhood home. Mom made coffee and tea and served homemade French bread, with powdered sugar on it. We kids just smiled at each other. I looked around the house with the its comfortable living room, the pictures on the wall, my father's thick books on the bookshelf, the potted plants on the window sill, and the tiled stove. Then there was the view from the windows – I stood there, looking at it. Victoria and Hans followed me, laughing, into the kitchen, looking at the warm stove. I grabbed the ring in the floor, and tried to open the door to the basement. Then we went upstairs to the bedroom, to look out at Store Marlene from the window. How nice it was to be home. 'Let me stay here, let me stay here', I thought. We three siblings sat down on Mom's bed. I touched my big sister's long thick braids, and she pulled at my short 'Prince Valiant' hairstyle, and my little brother tickled me. We had a lot of fun, until Mom called us down for tea and French bread. On my way down the stairs, I sat down and thought about the times when we little ones had sat there, eating our snacks, enjoying the views onto the mountains. Mom was all smiles. 'Eat', she said, pointing at the food, with its brown sugar. I thanked her, and she laughed with me. Suddenly, the hour had passed, and Miss Ingrid told me that it was time to go home. 'Home? This is my home', I thought, but I had to go back, reluctantly, to the orphanage. That night, in despair, I cried myself to sleep again.

New Skis

One day, we were asked to stand next to one another, all in a long line. Standing there in front of Miss Blom, we all had to stretch out our right arms above our heads and bend our fingers. 'What size shoes do you wear?' we were asked, whilst we were being measured. All of our measurements were being written down, next to our names. Skis, ski poles, and ski boots were going to be donated to us. Finally, the skis arrived. On Sunday, we were going to go on our first ski trip, but first we had to wax the underside of the skis, using an old candle, in order to make them run more smoothly. We were told repeatedly, 'Make sure you get the groove on the boots where the bindings fasten'. Even Benze had been skiing before. 'Can such an old wife really ski?' I thought to myself. We went over to the slopes near the football pitch, where we practiced, laughed, fell over, and got back up again – what fun it was! When we were putting the skis back in the basement, we had to first hold the skis together, tie them to one another at the top and the bottom, and then put a wooden block between the skis in the middle. Then we had to hang our ski poles on top of the skis. To begin with, we were only allowed to ski around the orphanage, where the adults could see us. But as we became more experienced, we could go over to the other side of the nearest mountains. At first, we went in small groups, accompanied by an adult, but later on, we could go all the way over to Lille Malene and ski. When we went to bed in the evenings, it felt as if we still had our skis on. Skiing for the whole day was glorious, especially when it was sunny.

One day, I was skiing downhill a little too quickly, on my way home from Little Malene. I hit a discarded bucket, and when I got up again, completely confused, I found I'd broken my right ski. It took me a long time to get home. Some of the girls kindly accompanied me, but others had gone ahead, spreading the gossip about my broken ski before I could get home. So when I finally arrived, a whole delegation was standing there. When Benze arrived, she examined the damage and then loudly announced that I would have to save up myself in order to replace the skis. In the silence that followed, I cried, thinking, 'How can I save up enough money for a pair of skis?' Benze said, 'Now, it's time to eat', and hurried inside.

I got home from school one day, happy, although I couldn't find the other girls. Up in one of the rooms, I could hear them giggling. I stormed into the room. When they saw me, they said that I had to get out, all the while hiding something behind their backs. I protested, loudly, but it made no difference. I got very upset and went down to Benze to report the incident, and she said that she would investigate. It turned out that they had arranged with Miss Blom to meet in a secret group, in order to knit birthday presents for me. I was the first to have a birthday in the new year. I thought that it was taking them a long time to make the presents, so my emotions were mixed; whilst I was excited about the presents, I was sad not to have anyone to play with. Finally, April 21 came round. The sun was shining, and the snowdrifts were smaller. 'Hooray! Congratulations!' came the greetings, from all directions. I had to wait until after school to open my presents; the table was covered with the white tablecloth, and on it were buns, sandwich cake, and hot cocoa. The other girls were just as excited as I was when I was unwrapping my presents. They had knitted clothes for my china doll, Tove. They sung birthday songs to me, and all the girls wanted to play with me. What a lovely afternoon! My doll was dressed so nicely. But I felt I was missing something. I was really upset that night when I went to bed, because I couldn't understand why my mother and siblings didn't visit me on my birthday. Even just for cocoa? There were no gifts from them, either.

On May 4, it was Albert's birthday, and May was a really good month, because six of the children had their birthdays – on May 5, Agnethe; on 11th Gâba; on 20th, Bodil; and on 24th, the twins, Eva and Marie. The following month, Ane Sofie had her birthday on June 30, and on July 30, it was Little Kristine's turn.

Good Enough for a Visit from the Queen

Little Kristine once told me, just after a ship arrived with supplies from Denmark, that Benze had ordered a whole bag of carrots. The same night that they arrived, she was going into town with some girlfriends, and Benze always had an admonition ready for when she was leaving the house. That night, the last thing she said to us was, 'Don't go into the basement stealing the carrots'. And to top it off, she added, 'That's told them'.

All of the snow had melted, and now we could be out playing until eight o'clock in the evening. Ordinary houses were being built all the way up to the orphanage. At the end of the road, they were building the new radio station, so stacks of building materials were lying around all over the place. It was great place to play hide-and-seek. It was great fun, when we were all together. '1, 2, 3, we've found Regine!' we shouted as loud as we could. We had found everyone apart from Sâmo. Where on Earth was he? In the end, we shouted out, 'Sâmo, Sâmo, come out now', but still he didn't come out. Eventually, we found him. Sâmo was lying, quite stiff, underneath a tall stack of boarding, and not uttering a sound. Pieces of bone were protruding from the upper side of his forearm – his arm was truly broken. We all dashed over to him, and Big Kristine ran for help. Regine was stroking his hair, and they whispered together in Greenlandic. I think she was trying to comfort him. As soon as Benze came, she looked at the injury, sent Miss Ingrid to fetch bandages, and wrapped up the protruding bones. Then she helped Sâmo to his feet, before going up to the orphanage, with all of us following her. Benze telephoned the hospital, and a car came for Sâmo and Benze. Poor Sâmo. When he got home, he had plaster cast on his arm, from his wrist to his elbow. We girls became very caring towards him; I think that we managed to teach him a lot of Danish during that time.

A few months later, there was an embarrassing silence at dinner time. It was past 6 p.m., and we were still waiting for the last of us to arrive home. Finally, Sâmo, Karl-Otto, and Helge came in at quarter past, each with blackened fingers, and dirty faces. 'Where have you been, and what do think you look like?' Benze asked. All three of them stood in front of Benze, and she repeated the question. The boys were whispering and pointing. Benze repeated her question again, and the boys carried on, whispering their answers. 'You know that you have to be home for dinner. The rest of us have been waiting for you. Now, you're grounded for a week'. That meant that they weren't allowed out of the house for a whole week. 'Now, go and wash your hands and faces', Benze ordered. 'Oh, no', I thought. 'They've asked one of the *kiffaks* to help them get some booze'. 'Enjoy your meal!', Benze said. It turned out that the three boys had been down at the site where the new Adventist church was to be built, rooting around amongst the tar barrels.

Spring meant the big clean-up, and a huge amount of washing. Once the bedding had been removed, the boys were supposed to help carry the mattresses out and beat out the dust with a carpet beater. We girls had to get into all the nooks and crannies, hand-cleaning all of the surfaces with a cloth, including the areas underneath each of the beds. We were grumbling and complaining, some of which Benze heard when she was passing. She stuck her head into the room, and said to us, 'It must be good enough for a visit from the Queen!', and then she went on her way, giggling. That remark worked, because we had met the Queen before, and we had talked about the possibilities of meeting her again. Luckily, we only had to clean our own rooms. Outside Benze's living room, there were long clotheslines, where all of the freshly washed linen hung in the wind. When it dried, it was ironed and folded up. When we went to bed that evening, it was a treat to have clean, freshly ironed bed linen to fall asleep in.

Benze dragged the big plant pots up from the basement and placed them on window sills in all the rooms. She planted tomatoes, chives, and radishes. We followed their development from day to day, and I remember that there wasn't much growth during the first part of the summer. A new job for us to do was to peel the new, smaller leaves from the side stems of the tomato plants, so that the tomato plants would grow taller and stronger. We already had cactus plants in the windowsills. They had thick fleshy lobes, and spikes, which stuck out like medical needles. This gave us an idea. Inside the medicine cabinet, there were large needles, and there were syringes in the drawer. We had seen how the needles were fixed to the syringes. 'Maybe we can find a way to get water into the cactus leaves?' So we tried to do this. We went into the bathroom to draw water into the syringes, then went quickly back to the bedroom, and inserted the needles into the lobes of the cacti. We all took turns; it was good fun, and the lobes did thicken. Afterwards, we cleared up, and went out to play. A few days later, some of the bigger girls had become even braver. They picked up the syringes with the needles, and asked, 'Who dares do this?' They placed the needle in the inside of the elbow joint, and we would see who dared to bend their arm furthest inwards. We watched with great interest. Suddenly, Bodil started bleeding. We all held our breath. 'What are you doing?' asked Miss Blom, as she came into the room. We had been found out!. Miss Blom went straight down to Benze and showed her the used medical needles. 'Tell-tale!' we whispered behind her. We got a proper reprimand and were grounded for three days. From then on, the medicine cabinet was locked. 'We've got nothing to do', we complained to all the others, at dinner.

'Cheep, cheep!' came the sound from the hollow part of the stone surround of the lawn. We went closer to the sound, in order to find out exactly where it was coming from and saw some young snow sparrows, who were waiting, open-mouthed, for their food. We were silent, because now we had a secret that, for God's sake, we had to keep from the boys. They had made themselves catapults, which they used to shoot stones at the sweet snow sparrows. And now it turned out that there were young ones, right here with us. Oh, how exciting it was. We girls ran around happily with our secret for weeks, until one day, when we got out of school. We could hear the boys shouting and whistling, and shooting stones at snow sparrows. The boys howled in triumph whenever they managed to hit one. We saw the poor mother snow sparrow. She was dead; she wasn't moving at all. With tears in our eyes, we scolded the boys, with honour and glory. We were now deadly enemies, and we girls chased the boys away from the dead snow sparrow. Some of us stood guard over her, whilst others looked for sticks that we could use to make a cross. We also found a small container, so we dug a small hollow, where we carefully placed the snow sparrow on its back. Then we started the funeral ritual. The boys returned, curiously, and we were reunited again. One of the girls played the part of a priest; we found a couple of hymn books, and we sang, 'In the East, the sun rises'. The 'priest' filled in the 'grave', and placed the cross in the ground next to it, and eventually we said the Lord's Prayer together. Over the summer, we did this a few times. We took it in turns to play the 'priest'. We sometimes sang a song that we had just learned; it was one about a little girl who was dying and singing a farewell song to her mother. It was so sad – 'Mother, I'm tired, now I want to sleep'.[28]

'Straighten your back'. 'Pick up your feet'. We did not lack any reproaches, big or small, or commands as to how we should act. It was hard to do everything well enough. If we lost a button, we would have to sew a new one immediately. If our socks had holes, we would have to stop what we were doing immediately, and darn them ourselves. This struck one to the core – at least, it did with me. I was very dutiful. One day, Benze had had enough of those who couldn't remember how to do what she had been trying to teach us. We girls were called together in the hallway, down in front of the stairs. 'Line up. Not you Agnethe and Helene, you stand next to me'. Regine, Ane Sofie, Bodil, Julie, Big Kristine, and Little Kristine were told to stand facing us. First, Benze pointed at their missing buttons. Then she pointed at us. 'Can you see the difference?' she asked, sharply. 'Yes', the others whispered. I stood there, my toes curling; I didn't think that Benze could be that familiar. 'Tomorrow, everyone who has buttons to sew on, and socks to darn, will have to do so as soon as you get home from school. Then we'll see if you get to go out playing before you do your homework'. Agnethe and I got the 'look that could kill' from the other girls. Amongst us girls, it was very quiet at supper.

'Your foster parents, Bishop and Mrs Fuglsang-Damgaard, are coming over on the next ship to visit you', Benze told Barselaj on breakfast time. We pricked up our ears and looked around at each other, especially at Barselaj. How lucky he was. We had come to miss our foster parents. 'They must be rich, if they can travel to Greenland just like that', I thought. They were beautifully dressed when they came and ate with us. They had a brand new bike for Barselaj. The lady also brought a gift for all of us. We girls each got brooches, with beautiful embedded stones, which were shaped like small guitars. We were very happy and sat for a long time, just looking at them. One by one we went up and thanked her, and shook her hand. I don't remember what the boys got. There were now two of us at the orphanage who had bicycles. We would cycle from the orphanage up to the dam at the new reservoir, down to the end of the football pitch, and back home. I let the girls from my room take it in turns to borrow my bike, and the other girls started asking if they could, too. 'You could charge people 25 øre to borrow your bike', said Big Kristine. I thought it was a good idea. Barselaj didn't want to let people borrow his bike, but when he saw that I was getting money for it, he agreed to the idea. So we earned some extra pocket money, and I was able to afford to buy glossy pictures in Ole's Department Store. Eventually, I could afford a picture album; then I could swap glossy pictures with the Danish girls in my class.

It was pocket money day, so we rushed home from school. We lined up in front of Benze. Once we got the money, our names were ticked off on the list. 'Who wants to come with me to KGH?' Four of us girls set off there, and stood in line to ask for a bar of *Tommelise* chocolate and a packet of *Mariekik* biscuits. We could get that for 75 cents. Happily, we went up to the top of the nearest hill and sat down. We enjoyed the view, whilst alternately putting pieces of the chocolate, and bite-size pieces of the biscuits, into our mouths. They were just so delicious. One of the girls decided to put the biscuit between her front teeth and to gnaw it down gradually from the outside edge, so of course, we all had to try that. How we laughed – it was such a pleasure, absolutely wonderful. We stored half of the biscuits and the chocolate in our cupboards, saving it for the next day.

In the bright summer evenings, we often played rounders or stick-ball in front of the orphanage. The oldest children were always either the first or second to

FIGURE 16   Barselaj with his bicycle

be chosen for the teams; naturally, the youngest were chosen last. It was only rarely that they would last for more than a few rounds. We who lived at the orphanage were used to the everyday harshness of Benze's tone – for example, when someone did not treat the toys properly. Then she promptly called out the offenders. You would be sent to your room, until she pronounced you forgiven. If we wanted our classmates from the Danish school to come home and play, we would have to make such a request no later than a day in advance.

Notes

1 DRK stands for '*Danske Røde Kors*' – the Danish Red Cross.
2 Both are red-and-white flags, with crosses – the Danish (and International) Red Cross flag is a small red cross on a white field, and the Dannebrog (Danish national flag) is a white Nordic cross on a red field.
3 A *kiffak* was generally a Greenlandic woman, employed by Danes, in the position of a poorly paid domestic servant. Given its specificity, I have left this word untranslated throughout, although I have pluralised it using English rather than Danish conventions (i.e., '*kiffaks*', rather than '*kiffakker*').
4 '*Trusse*' is a Danish word for female underwear, and the suffix '*ly*' appears in many place names. Hence, a rough (American) English translation of '*Trussely*' might be 'Pantyville'.
5 Before the more widespread use of the character 'å', the corresponding vowel sound was often represented as 'aa' in written Danish (and Norwegian). This convention persists in some older Danish and Norwegian surnames (e.g., 'Nygaard'), and here, in older spellings ('Godthaab' instead of 'Godthåb').

6 In the Nordic countries, more so in years past, occupations are/were often used in appellations, especially in making introductions; at least, to a greater extent than are, or has been, the case in Anglophone countries.

7 'Asmussen & Weber' and 'Højgaard & Schultz' are the names of local companies.

8 Somewhat related to endnote 6 (above), at this time in the Nordic countries, the holding of a Master's degree (here, Meldgaard), or studying for one (here, Parbøl), was reflected in a person's appellation.

9 Literally, 'red groats', or 'red porridge', *rødgrød* is a sweet fruit dish, popular in Denmark and northern Germany.

10 A well-known traditional Danish hymn, composed by Christoph Ernst Friedrich Weyse in 1837. Its title translates literally as 'The Sun Rises in the East'.

11 In Danish, rosefish (also known as Atlantic redfish, red perch, or red bream) is '*rødfisk*' – literally, 'red fish'.

12 This is the Danish translation of Helen Bannerman's '*Little Black Sambo*', a book read to and by generations of children around the world since it first appeared in 1899, but obviously less often in recent decades.

13 Literally, 'I am tired, and going to sleep' – a traditional Danish Christian 'goodnight' song for children.

14 The titular character of an 1845 literary fairy tale by the Danish writer, Hans Christian Andersen.

15 There is not really an appropriate English language equivalent of the Danish *råkost*, so I have left this word untranslated. It is a mixture of uncooked fruit, salad, and root vegetables; the idea is that when these ingredients are combined properly, all (or at least, most) of the essential vitamins and minerals are contained.

16 Literally, 'Peter's Christmas', a Christmas story written by the Danish historian Johan Krohn, which first appeared in 1866. For many Danes, reading this story is part of a traditional Christmas – an English-language equivalent might be Dickens' 'A Christmas Carol'.

17 *Juleherter* (literally, 'Christmas hearts') are traditional Scandinavian Christmas decorations, made by interweaving two heart-shaped pieces of coloured paper (usually, red and white).

18 In the Nordic countries, most of the Christmas celebrations (e.g., Christmas dinner, and the unwrapping of presents) take place on the evening of December 24, rather than on December 25.

19 In the Nordic countries, a *mandelgave* (literally, 'almond present') is given to the person who finds the almond that is hidden in the rice pudding that is traditionally served at Christmas.

20 *Risalamande* (Danish rice pudding with cream and almonds) is a dessert, traditionally served after Christmas dinner. If served in individual portions, one of them will contain the whole almond that entitles them to the *mandelgave* (see above, endnote 19).

21 *Far til Fire* (English title, 'Father of Four'), is a 1953 Danish family comedy film, based on a popular comic strip. It was the first in a series of eight such films, which were released annually between 1953 and 1961.

22 The Greenlandic form of 'Helene'.

23 The correct Danish here is '*Klokken er fire*'.

24 This is a joke that is lost in translation, as is the reason why the young Helene's mishearing was wholly understandable. In Danish – especially in certain accents – the word for 'Indochina' ('*Indokina*') is a near-homophone for the phrase 'someone you know' ('*En du kender*').

25 *Billedbladet* is a weekly celebrity and entertainment magazine in Denmark, which first appeared in 1938. Its circulation in the 1950s was approximately 150,000.

26 *Fastelavn* (Shrovetide) is a pre-Lent festival, celebrated in some of the Lutheran countries of Europe. Traditionally, the celebrations include the game of *slå katten af tønden* ('hit the cat out of the barrel'), which is much like *piñata;* the (hopefully gently!)

flogging of one's parents out of bed in the morning, using the decorated *fastelavns-riser* (Shrovetide Rods); and the much-looked-forward-to *fastelavnsboller* (Shrovetide cakes), which are small, round, sweet iced buns, often filled with cream. The lyrics of the children's song that Helene learnt to sing here contain a (humorous) threat that the children will cause trouble if they don't get their Shrovetide cakes.

27 Lofoten is an archipelago off the west coast of Norway, which has been a centre for cod fishing for over a thousand years.

28 This is a song based on a poem of 1830 by Hans Christian Andersen, entitled '*Der døende barn*' ('The Dying Child').

5

1953–1956

Greetings from the Greenlandic Children!

Via the journal of 'Save the Children', Benze sent a joint letter to our foster parents in Denmark:

'Greetings from the Greenlandic children!

Save the Children's sixteen Greenlandic children returned to Greenland at the end of September last year, after spending eighteen months with you. Miss Bengtzen, the Mistress at the Danish Red Cross's orphanage in Godthåb, where the children are now living, sends her greetings to their foster parents. The children and I have agreed that through this joint letter, we can send all of the foster parents our greetings.

It was with some sense of trepidation that I went aboard the ship Umanak, with the children. The journey from Copenhagen to Greenland is not exactly a pleasant one, especially if you aren't a good sailor. But regardless of any initial worries, the journey went well. During the first few days the sea was a bit rough, so some of us had to have to look to the sea gods for help. But we had a good time on board the Umanak. The cabins were nice, and the captain and serving crew were good to us – the journey exceeded our expectations. The children slept well, and ate and played all day long; in fact, they were a little bit spoiled. We had many 'goodies' with us, for which we had the foster parents to thank. Sometimes, the children would get up at between five and six o' clock in the morning. "Miss, isn't it the time that we get our chewing gum now?" One day, a children's movie was shown, so we didn't lack for any entertainment, either.

Umanak sailed through the Prince Christian Sound, which is a truly beautiful seaway, so we avoided having to round Cape Farewell.[1] The

DOI: 10.4324/9781003241843-7

views were magnificent, with high mountains on both sides of us. The children all stood out on deck, enjoying the trip. We arrived at Julianehåb late one night, and the following day, the children and I went for a long walk in the mountains, and picked blackberries. Eli was up at the Sanatorium, and greeted the little friends, and Miss Villadsen. We sailed on, arriving at Narssaq in the glorious sunshine, and we went ashore here, too. We saw the factory, and the sheep slaughterhouse, where we bought some lamb, so that we could be sure that we would have something good to eat when we got to Godthåb. At each stop-off point, the children asked, "Are we going to stay here?" But we sailed on, and arrived at Godthåb in the wind and rain. Many people had come to the dock, in order to welcome us.

We travelled by bus to the orphanage, and the flag had been raised – Miss Blom and Miss Holm were there to welcome us. It is a lovely orphanage, on top of a hill, with brightly-painted walls, and it's a good place for the children to play. The children said, "We want to stay here now, it's nice". During the sailing, the children had all got along well together; however, when we arrived, there was a little incident. Some of them had fought with some other children, and a bypasser had asked, "Do you need any help?" "No thanks", came the reply, "We can stand up for ourselves, because we all stick together".

The children now attend the Danish school, and they're doing well. Some of them receive extra tuition at home. The children are all healthy, and they often talk about their foster-parents in Denmark. There is no doubt that they have seen and learned a good deal, that they remember now, and will benefit from in the future. They still use bits of the Greenlandic language, but for the sake of the school, we must stick to Danish.

Every day now we have frost, snow and sunshine. The children are rolling around in the snow, building snow caves, tobogganing and skiing. The ski slopes are around the house, so that we can watch out for them from the windows. Sometimes, we have to use the bulldozer to clear the snow, so that we can get water, oil and food up to the orphanage.

Finally, all of the children send many loving greetings to their foster parents.

With kind regards,
D. Bengtzen'.

You Dummy, You Can Hardly Speak Danish!

Our school friends were from the Danish School. The school was split into two parts; one half was the Danish School and the other half was the Greenlandic School. We shared a playground, but even there, the two 'halves' didn't play together – we had nothing to do with the Greenlanders. I can't even remember if I ever had any contact with my siblings in the schoolyard. We couldn't communicate; it was as if we were invisible to one another, both when running around in the schoolyard, or playing on the hill in front of the vicarage, where we sometimes sat down on the rocks, swapping our glossy pictures.

Sitting there in pairs, our school furniture looked stark. The teacher's table was black, and the bottom parts of the windows were painted white, so that we wouldn't get distracted through looking out of them. We got vitamin pills at school – a small green pill, vitamin A, and a vitamin D pill. Occasionally, a large container of grated cabbage, carrots, and apples was delivered to the classroom. I had to be quick, if I wanted some. Our first dance teacher was Benze's friend, Miss Petersen, who was known as 'Petter', and was tough. Her blonde hair was set in a bun at the nape of her neck, with a wave over her forehead. She wore large glasses and usually dressed in a blouse and a pleated skirt. For many years, our principal was Knud Binzer. His son, Lasse, was in our class. We had other teachers with 'Danish' children over the years; the ones that I remember the best were Helge Lorenzen, Henning Forchhammer, Lotte and Hemming Hartmann Petersen, Else Andersen (who was the librarian), Grethe Bågø Nielsen, Kaja and Niels Toubro, Knud Poulsen, and Knud Breum Rasmussen. One of my old school friends, Margaret, contacted me after the release of '*I den bedste mening*'[2] and had told the author:

> I remember some of my classmates clearly, especially Eva and Marie, who I often played with. In class, I sat next to Agnethe Titussen, and I never forgot that her birthday was on May 5th. I also knew Big Kristine very well, and Ane Sofie and Ole Fly, who were really good at drawing, and were lots of fun to be with. And then I remember Eli, Barselaj, Little Kristine, and you. I often went to birthday parties at the orphanage, and the children from the orphanage came to my birthday parties. In those days, you always wore the princess dresses, and we were a bit envious of them. Did you know that? We had no idea why you were at the orphanage. Mom asked me why Eva and Marie lived at the orphanage, when their mother lived in town. We didn't know why, and Eva and Marie couldn't explain it, either. My father knew their stepfather, Karl Sildepisker. They also had a big sister named Rosa, who had a blood disease. It's a bit strange that I've never thought of you as having had problems in being "different" at that time. I don't think that we Danish children thought about, or even knew, that you couldn't speak Greenlandic, much less about why that would be a problem. After all, we couldn't, so I just thought that we were in the same boat. We didn't feel any different from you. Maybe the posh kids did, but Birte and me, who were different because we weren't posh, we certainly didn't. But when, as an adult, I read about how you were messed around, I understand just how unreasonably you've been treated. At the time, it seemed to us that you were "luckier" than other people. You lived in the orphanage, and had a lot of fun, and togetherness – and a lot of toys. We didn't have that.

Whilst we always had to ask permission to bring friends home a day in advance, we didn't have any fixed visiting times. In the fourth and fifth grades, Margaret came over to play with Ole and Eli several times. The friends we invited experienced Benze's changing moods. Sometimes, she could be sweet and caring, and on those good days, we'd draw, or crochet, or play together. But on her bad days, she would yell at us, and sometimes lock us in our rooms, and say to our guests, 'It's time for you to go home'. Several people we invited back to the orphanage told us about sitting with us in the big kitchen, enjoying ourselves, but feeling scared when Benze came in.

Vivi and her big brother, Ejvind, came over to the orphanage to play quite often. Their maternal grandfather, Jørgen Chemnitz, was well-known locally. During the royal visit in 1952, Vivi was seven years old, and this grandfather served as an interpreter for Queen Ingrid and King Frederick. When the royal couple were being shown around the town of Godthåb, a crowd of children and adults were following along. Vivi noticed her grandfather, and shouted, '*Ittu! Ittu!*' ['Grandpa! Grandpa!']. Smiling, King Frederik asked Jørgen Chemnitz, 'Who's that?' Chemnitz replied, 'That's my granddaughter'. After that, King Frederik put *Viviaraq* [little Viv] on his lap for the rest of the tour. Vivi told me that her mother taught Queen Ingrid a lot about the Greenlandic community. Vivi has shown me the longstanding, very friendly correspondence between Guldborg Chemnitz and the Queen. The two families also exchanged photographs. Vivi clearly remembers that a great deal of preparations were made before the royal couple arrived, in the summer of 1952. Greenlandic national costumes were made for the three little princesses, and her maternal grandmother had helped to sew them. Vivi served as the model for Princess Anne-Marie's national costume. Vivi was also busy during the municipal elections, where she went out delivering posters with her maternal grandfather. She was yelling out, 'Vote for my *ittu*! Vote for my *ittu*!'. She was running around everywhere, and fell over, and ended up with a scar as a keepsake.

One time, my second cousin Ane-Johanne was looking after Doctor Rendal's son, Bent, and Regine and some of the other children from the orphanage were over there, playing. Regine walked right up to Ane-Johanne, and shouted straight into her face, 'You dummy, you can hardly speak Danish! You dummy, you can hardly speak Daaa-nish!' When the local people met us orphanage children in the mountains, they didn't want to be near us. We were told this, both outside, and at home. The locals seemed to think of us as living on our own little island. This was very strange, because Ane-Johanne, who was, as I said, my second cousin, was Agnethe's best friend. Before my father died, we had always played together. But now, this was forbidden.

One day, we had seen the boys having a pee-ing contest from the edge of a cliff. We were not impressed. Led by the oldest girls, Big Kristine, Julie, and the twins Eva and Marie, we decided to have our own pee-ing contest – not on the hill facing the orphanage and the town, but rather on the other side of the mountain, facing the harbour. We found a good rocky ledge, where there was enough space for all of us. We got ready, all standing in a row, pulled down our brown underwear, with the elastic in the legs, and got down on our knees. We were laughing at each other when the twins said, 'One, two, three, pee!', and we released a whole series of pee-streams. We were so preoccupied with this activity that we had not seen Eli and Ejvind, standing downhill, completely still, watching what we were doing. We only noticed them when they started to laugh. 'Don't laugh!' we cried out together, a bit embarrassed at having been exposed.

Vivi didn't only play with us girls; in fact, because she was a real 'tomboy', she played with the boys more often. One day, she had Albert and Barselaj over to play at her house. Amongst other games, they had been playing hide-and-seek, and she had locked Albert in the coalhouse and shut Barselaj inside a travel trunk. After they had finished playing, and it was time to go home, Vivi's grandmother was sitting in the living room, having her coffee. Suddenly, she heard a knocking, coming from inside the trunk. Although she was frightened at first, she went over to the trunk and opened the lid. She got another shock when Barselaj popped up, looking exhausted. She asked who had shut him up in the trunk. He explained that they had been playing hide-and-seek and that Vivi had closed the lid. Vivi's grandmother helped Barselaj out of the trunk, and then he ran home. '*Viviaraq*, how could you even think of shutting poor Barselaj up in the trunk?' she asked. Vivi stutteringly explained that she did not know that the trunk locked automatically when the lid was closed and that they had started a new game, and had completely forgotten about Barselaj. Her grandmother scolded her: 'You must never lock each other up in things, do you understand? And now you're grounded for ten days!'.

Those of us orphanage children who came from Godthåb obviously had cousins in the town, but Benze didn't care about that. Some of our aunts made inquiries, mostly through the *kiffaks* who worked for Benze, as to whether these cousins could come and visit us. They were told that this was not possible, because they were Greenlanders. When Vivi's mother heard about this, as a government official, she tried to 'pour oil on troubled waters'. That didn't achieve anything. She was very upset that she couldn't help, but on the whole, the children in the town now seemed strange to us. They admired us because we could speak Danish fluently, and they envied us because we lived in the orphanage. We radiated perfection and health, always clean and neat, in nice clothes, with nice hairstyles, looking happy and smiling. As we walked through the town, we might take it in turns to choose one of the songs that we had learned, or the boys would race each other through the streets. But of course, not everyone admired us. Some children, Edward and Judithe, lived near the orphanage. They were purely Greenlandic-speaking. The boys were enemies with Edward, and we girls were enemies with Judithe. We shouted ugly words at each other, in the different languages that we spoke. The boys knew from doing sports together that Edward didn't always wear underpants. When he had shouted something incomprehensible to them, they shouted back at him, 'Underpants! Underpants!' When we passed the house where Edward and Judithe lived, it was usually on our way home from school. So it wasn't any fun to accidentally meet them in the town. One time, I was inside Echwald's bakery shop, buying a ten-øre cake. When I was leaving, Edward, Judithe and some of their friends were waiting outside the bakery. They started shouting something in Greenlandic at me. I pretended that I hadn't heard, and tried to slip past them, but one of them picked up a stone and threw it at me, and it hit me right on the cheekbone. It hurt, and they suddenly looked frightened. Inside the KGH store, someone had seen what had happened and was rushing out, ready to chastise them. 'Don't walk home alone again', the other girls said, when I was answering their questions about what had happened. This comforted me, but when I got home, I shut myself in the basement and cried.

We used to refer to the janitor as 'Crappy Jens'.[3] That nickname came from the time that he used to have to physically empty the toilet bowls, before we got flushing toilets at the orphanage. He had a daughter, Naja. She was a cousin of my mother's, but I didn't find that out until many years later. Naja and her husband were the first people to build a house on the land below the orphanage. They moved there in 1953, and Naja still lives there. Her father told her that many, many children lived up at the orphanage, who got good food and were very spoiled, but he couldn't speak to them. He had wanted to speak with me, of course, because he knew that we were related to one another. My only thought had been that he was very nice.

During the summertime, we would make full use of the large terrace. To make the place in front of the terrace as flat as possible, large pieces of broken rock had been moved and used in constructing the tall, black tarred plinth, and the cement staircase. After Benze had sown a lawn, we could hang our blankets from the terrace, and make 'tents'. It was so nice – we brought books, pillows and dolls into the 'tents'. During the autumn, Jens would get ready to hang the freshly slaughtered lambs up on the racks. It was great when we managed to get some tallow from the bellies of the lambs; we had thought the adults would forbid us to do this if they found out about it, but instead, the opposite happened – they took pictures of us whilst we were doing this. Of course, as sheep, rams and lambs roamed freely around Godthåb, and they grazed around the orphanage. When the sheep had their lambs, we used to run after them, and try to feed and pet them. The boys liked to tease the ram. 'Stop doing that!' we girls shouted at them.

One lovely summer day, the boys lured the big ram onto the steps that led up to the terrace. 'Baaa! Baaa!', they said, as they walked backwards up the stairs. They lured it up there with a newspaper, which they were waving right in front of them. We watched, to see what would happen. Just in front of the open terrace door, the ram chomped at the newspaper, and lunged violently towards them; suddenly, he was inside the dining room. The floor was very smooth, and the ram looked like Bambi on ice. But he was a bit clumsier than Bambi was. There were shouts and screams, and the knocking over of chairs and tables. Slowly at first, many of us children ventured inside the dining room. What a mess! We had closed the doors that led to the hallway and to the kitchen, but suddenly Karen and Benze came rushing in. 'What are you doing?' they yelled. Lots of us jumped on top of the tables, for safety. On the inside, we found it highly amusing, but on the outside, we looked a bit scared. Eventually, Karen and Benze got the ram out through the terrace door again – which was also fun to watch. But it wasn't so much fun afterwards. Everyone who had been there at the time was grounded for two days. We were really mad at the boys – after all, it wasn't us who had lured the ram inside.

Kapisillit

During the summer of 1953, all of the adults and the children from the orphanage went on holiday together for the first time. We were certainly looking forward to it. Benze had good connections, so she booked us on the medical ship, *Saxtorph*, to sail up the fjord to Kapisillit, and back again. The journey was one of more than one hundred kilometres and would take several hours. The lead-up time to our departure was exciting. We went down to the harbour when *Saxtorph* was docked there, to see what sort of a ship she was. Just standing there, smelling the sea, and hearing the waves, was enough to raise our expectations for the summer holidays. The wait seemed endless. For weeks, the house was buzzing with all of the hustle and bustle. We used big wooden crates to pack all of the things that we would need. First, we packed the long-lasting foods – oatmeal, potatoes, sugar, and so on – and we packed the bed linen, towels, and washcloths in another crate. We packed the perishable food items the day before we sailed. We also had to pack our own clothes and provisions, including the white mosquito nets that had been made for each of us.

Finally, the day of our departure arrived. As I said before, the beautiful Miss Ingrid came from Kapisillit, and her father was the supplies manager there. The sun was shining, and our excitement was reaching its peak. We walked arm-in-arm, with our jackets open, down towards the colony harbour. Our little suitcases had been driven down there already, along with the mattresses and blankets. All of our boxes and packages were everywhere around the ship, even on the deck. At last, we were ready, and the thick rope attaching the ship to the dock was loosened. There was a splash when the rope hit the water. 'Well, that's the start of things', I thought, as one of the men in oil-spattered clothing pulled up the wet rope, and secured it to a thick iron ring. Then, I heard the distinctive sound of the engine starting up. 'Oh yes, now we're sailing!' The colony harbour was getting smaller and smaller, as the ship turned towards the right. I saw my grandfather's little white house in the Islandsdalen, and the twins, Eva and Marie, pointed towards the shore and shouted above the noise of the engine, 'Our mother lives there!' They were pointing at one of the first houses that had been built in Myggedalen. Then, we sailed around another bend and got a good view of Godthåb's famous mountain, *Sermitsiaq* (or *Sadlen* ['the Saddle'] in Danish). How beautiful it was. Some of us ran all the way up to the bows, and sat at the edge, with our legs dangling over the side of the ship. We held onto the rail, and we started singing 'We're sailing up the river, and we'll be sailing back down again' at the tops of our voices. It seemed like a great song at the time. Oh, how wonderful it all was.

We looked closely at all of the beautiful mountains, taking in every one of their features that we could. One of the children knew a spooky story about a rocky spur at the top of one of the mountains. Local legend had it that a man was forced to jump to his death from that point. We gasped at the thought of it. We sailed along the coast, and sometimes it looked to us as if we were going to sail straight into the mountains, but then the boat would make a turn, and we would get a long view towards the next piece of coastline. Eventually, we were all shouting, 'Look! There's Kapisillit!' at each other. When we could see all of the little wooden houses, painted in all their different colours, the place looked beautiful to us. When we saw a bright yellow house, with a number painted on the roof, Miss Ingrid said to us, 'My parents live there'. It was a big house. 'We should visit them, on one of the days'. As we got closer to the harbour, we could see that it was low tide – we could see the high tide marks on the thick wooden posts. It took a while for us to dock. We walked up the thin, round iron steps, that went vertically up the quayside – obviously, we had to wait for one another to do this. There were lots people on the dock, greeting us, and waving to Miss Ingrid. Several people tried to ask us things, but when we couldn't answer them, because we didn't understand what they'd said, they looked at us in astonishment. The embarrassment of not being able to understand or speak Greenlandic hit me again; but fortunately, I had to get over these feelings quickly, because now we had to carry our suitcases up to the brown buildings where we would be staying. The buildings were up on the mountainside, and the view of the fjord was breathtaking. So much of this place was beautiful. A little way further on there was a beach, and we could hardly wait until the next day, when the adults had promised us that we would be able to go to swimming there.

When we had unpacked, we were shown around Kapisillit, in groups. On the outskirts of Kapisillit, there were some big, tall drying racks, stocked with cod that was hanging out to dry. The fish had been sliced at the belly, their heads cut off, and they were tied together in pairs at the tails and hung on the rack. I was in Miss Ingrid's group, and she told us that it was her father who had invented the drying rack method and that now such racks could be seen in many places throughout Greenland. 'Why have the cod not been cut up into flat pieces, like the ones we're used to?' we asked. 'Because the cod that's dried in this way is sent to southern Europe, and they eat cod in a different way than we do'. There weren't that many houses in Kapisillit. We were shown where the dining room was, and where the outside toilets were, before we were finally allowed to run around freely. Most of us went down to the beach. There were lots of pebbles, and the water was like a mirror, so we made our wishes to the surface of the sea. Our bunk beds were cosy, but it was difficult to fall asleep on that bright summer night (Figure 17).

The next day, we went hiking in the mountains. Each of us brought their own packed lunches, and the *kiffaks* brought a Primus stove and a kettle, too. I thought that it was so beautiful – bounteous nature, as far as the eye could see. We ran and jumped through the mountains, in front of the adults, who shouted at us to 'Stop!' when we got too far ahead. At one point, we made camp. We

FIGURE 17 On the beach

were supposed to be helping to gather heather to build a campfire, but as soon as we stood still, the mosquitoes came buzzing along. We ran quickly towards the adults, in order to grab a mosquito net each. There were a good many of us, so we were able to gather the heather quickly, and whilst we were doing this, the *kiffaks* built a fire between two large stones. They then filled a large pot with water from the nearby river. We were happy to drink the river water; we used the hollows of our hands and slurped it up. It was great to be able to drink water that way. I thought for a moment about my family's summer holidays in Qoo-qqut, when my father was still alive. When the water in the pot over the heather fire had boiled, there was tea for everyone. The Primus stove was used to heat water for the adults' coffee, which they made in an old *Madam Blå*[4] coffee pot. We found a nice spot on the mountain to eat our packed lunches, but we found it difficult to get our food into our mouths, because the mosquito nets kept getting in the way. We laughed and tried to remember to lift the mosquito net up each time that we wanted to take a bite of bread.

The day after that was 'a big washing day'. The shout came, 'Put all of your clothes together, so you can find them again!', and then we had to go out into the water. Oh, how cold it was! There were lots of pebbles, too, but it was fun, and every time we walked out into the water, we went a little bit further out. The day ended with a visit to Miss Ingrid's mother and father, who had invited us for cocoa and buns. They had a nice, cosy home. Miss Ingrid's father told us about the delicious salmon we were going to have that night. 'In the summer, the salmon go up the big, clean river in Kapisillit, in order to spawn'. 'What do they do?' a few people asked. 'They fertilise their eggs. First of all, the female salmon come along, and shed their eggs, in long lines, in the river. Then, the male salmon come along, and deposits its seed on top of the eggs. This happens in the month of October. Then, the little eggs sit between the small stones of the riverbed before they hatch. The adult salmon overwinter in the river, before they go back out to sea in the spring. The young salmon stay in the rivers until they are about five years old. At that point, when they are about twenty centimetres in length, they too go to sea. Over the course of the next few years, the young salmon become sexually mature'. 'What do they become?' asked the boys, giggling as Miss Ingrid's father explained what that meant. He continued, 'Then, the salmon return to exactly the same river in which they were born, returning to one of the cleanest rivers in Greenland, namely Kapisillit'. We groaned. Just imagine, that's how Kapisillit had become famous, and now it was being invaded by recreational anglers. These were the salmon we were having for dinner. Wow, they tasted good, and it was great to eat a fish other than cod or rosefish. From the harbour, we could watch all of the motorboats and dinghies, and there was also a place where the boys could play football. All in all, our first summer vacation with the orphanage was a wonderful trip.

We could hear the melody of '*Den toppede høne*'[5] coming from inside the dining room. The door was closed, but we charged in. We were met by people placing their forefingers to their lips. It turned out that Barselaj and Bodil had started to take piano lessons. If we sat still and quiet, then we'd be allowed to watch. So that's what we did (Figure 18).

They were both talented and were always amongst the quickest to learn new songs. We sang a lot, both popular songs and hymns. '*Det døende barn*', a poem by H.C. Andersen, was one of our favourite songs[6]:

'Mother, I am tired, and now I want to sleep
Let me sleep next to your heart;
But do not cry, you must promise me that, first
For your tears burn my cheek
Here, it is cold, and outside, the storm is looming
But in my dreams, everything is so beautiful
And I can see the sweet, angelic children I see
When I shut my tired eyes.
Mother, do you see the angel by my side?

FIGURE 18 Piano lessons

Do you hear the lovely music?
See, he has two beautiful, white wings
The ones he was probably given by our Lord;
Greens and yellows and reds float in front of my eyes
They are the flowers which the angel is scattering!
Will I also get wings while I am alive
Or, mother, will I get them when I die?
Why are you holding my hands so tightly?
Why do you press your cheek against mine?
It feels wet, and yet it is burning like fire
Mother, I will always be yours!
But you must not sigh anymore
If you cry, I will cry with you
Oh, I am so tired! I must shut my eyes.
Look, Mother! The angel is kissing me now!'

That song could really bring up some strong emotions. The most sensitive ones amongst us always had their handkerchiefs ready when we sang it. In my longing for my mother, I felt as if I was in the same boat as the poor little girl. It was much the same with '*I en seng på hospitalet*',[7] a song that we also loved greatly.

In the middle of November, it was time to write Christmas cards to our Danish foster parents. We started well in advance, so that our Christmas cards would not be too late for the last ship that took mail to Denmark before Christmas. Benze was always on time with all things. People from the postcard company 'Forrania' took photographs of us for the Christmas postcard. On the back, we then had to draw Christmas motifs in the space where the address was usually written, and next to that, we had to draw out five lines, on which we were to write a message. At the top, on the right-hand side, we had to write 'Christmas 1953'. Here's what I wrote to my foster parents:

> Dear Mrs Greve and Mr Greve. A really Merry Christmas and a Happy New Year. Wishes from all of us here. The most loving Greetings from Your Helene.

Then I coloured in my drawings of garlands, a Christmas tree with many gifts underneath, three pixies, a snowman, and two little pixies in a cave. We were working away on this task, borrowing erasers and rulers from one another. Many of the children could not think of anything to write or draw, so they sat just there, sighing and yawning. I was one of the few who thought it was fun, and a nice thing to do. The ones who finished this task quickly were allowed to make Christmas decorations. Benze had to keep track of things; we had written our names in pencil, in the bottom right-hand corner, because not everyone finished on the same day. Afterwards, Benze had to find the addresses of our foster parents in Denmark, and our biological parents, who were in Greenland. She also sent them each a photograph postcard, which she wrote in Danish.

Out in the kitchen, Karen was buttering the freshly baked muffins, which we were going to have, along with a soda, whilst we were writing our Christmas cards. Benze went into the kitchen, to check if things were progressing well. Karen was really happy with us, so she was treating us by putting a generous coating of butter on the buns. Naturally, Benze saw this. 'You need to save the butter, Karen'. With lightning speed, but almost inexplicably, Karen replied to Benze, 'Sipaa, sipaa, sipaa – save, save, save'. Benze understood the broken Greenlandic/Danish words well, and she promptly dealt Karen a stinging slap. When Karen came in with the buns, we could sense that there had been some drama in the kitchen. One of Karen's cheeks was completely reddened, and Benze looked furious. She was used to the serving staff obeying her orders to save as much as possible.

Rabbits

One day, we were supposed to be going out hiking after school. We came home from school happy and went up to our rooms to change into our older clothes. But when we had got changed, Benze came into our rooms and told us, 'You can put on your nice clothes again. We're going to the opening of an old people's home, and you're to sing for the old people'. We complained, but with no success. When everyone was ready, we went along the school road to the new old people's home. I was walking hand-in-hand with Benze, and as we neared the old people's home, I asked her, 'What exactly is an old people's home?' She told me that it was a place where people went to live, when they got really old. I was angry that we were being dragged along there, so quick as a flash, I said, 'Well, you'll be able to move in there soon'. That made her angry. It was small comfort indeed that we got buns and lemonade when we arrived at the old people's home.

Next to the old people's home, they were building a brown, wooden building, on the mossy soil. It was going to be a new public bath. When it was built, after each gymnastics class, we had to go to there, along with the other children who were in the same school grade as us. A couple of women from the town looked after the baths; they were incredibly nosy, and always laughing together. It seemed to us that they were watching to see who was heading into puberty. When I started to get pubic hair, one of the women noticed it. I did everything that I could to try and stop them seeing, but one time, as I was drying myself, one of them started laughing and pointed towards my lap, saying (in Greelandic), '*Iliina tingeqalersimavoq*' ['Helene has pubic hair']. My face flushed with embarrassment, and I quickly pulled on my underwear and the rest of my clothes and ran back to class. I thought she was really stupid.

Every year, some of our classmates left, and we found that new ones had joined us when the school year started. Some of the older boys had a new male classmate, who knew a lot of bad language and crude songs. One day, we were hanging around, and the boys wanted to teach us some of these new words. 'Do you know what "banging" means?' We didn't. The boys were ready to show us and made an arrangement to do that the following day, over at the quarry. So the day after, three 'couples' went over there. 'Lie down on the ground', the boys said. We didn't really want to, but if we wanted to find out what the phrase meant, we had to do what they said. So we lay down, side by side. At first, the boys stood over us, but then they lay down on top of us, making the motions of sexual intercourse, as they breathed, groaned, and rubbed themselves. 'Get away from us!' we shouted at them. We girls agreed amongst ourselves that we could live quite happily without knowing what the new phrase meant.

One day, I heard someone shouting, 'Helene, you have a visitor!'. 'Me?' I asked, running down the staircase, where I met Mr Feldbo walking up. He was a neighbour of my first foster parents, the doctor's family in Copenhagen. Mr Feldbo had been very nice to me. He had taught me how to use a sledge and had taken me sledding a few times. I almost ran straight into his arms, but he offered me a handshake instead. 'What are you doing here?' I asked him, in surprise. He had found a new job in Godthåb, in radio. So he and his wife and children now lived nearby, in one of the newly built apartments. He told me that I was very welcome to come over and visit them, whenever I had the time. What a nice surprise! Mrs Feldbo was a homemaker, so she'd always be there. So now I had the possibility of a break from the orphanage, which would be lovely.

Another day, I heard someone shouting, 'Rabbits!'. Someone said, 'They have big ears, so don't make too much noise near them!' A man who was travelling from southern Greenland had given us two of them. The rabbits had to be kept in the bathroom, and it was mostly us girls who were interested in them. We got doll's beds, and doll's clothes, and dressed the rabbits up, and cuddled them – we were completely delighted with them. We gave them names – Petra and Petrea – and we arranged a 'christening' for them. We got some white fabric from the seamstress, who even helped us make baptismal gowns for the rabbits. At first, she didn't want to, but we managed to persuade her! We had a lovely few weeks with the rabbits, but one morning, when we went along to the bathroom, and they weren't there. We rushed out into the kitchen and told the adults the rabbits had gone. The adults looked thoughtful, but said nothing. Dinner smelled delicious that evening, and we sat down happily at the table, wondering what we were going to have to eat. But as we looked at the table dishes with the roasted meat, and then back at each other, we realised in horror that they had cooked our rabbits. We looked at each other, and shot each other glances that said it all – don't eat any of the meat. The adults tried to get us to eat, but I don't think that any of us ate anything apart from the potatoes that evening. After dinner, we went up to our rooms to talk. 'The grown-ups are evil'. 'Which of them thought up doing this?' 'I'll guarantee that it was Benze'. We talked about Benze for a long while, and we felt really angry with her.

Measles Epidemic

A good many people had fallen ill. There were students absent from every class, and we talked about this after school. Several of the children at the orphanage were confined to bed, too. 'It's measles', Benze said. 'What are measles?' we asked. But before Benze could answer us, the telephone rang. We tried to listen to what Benze was saying. 'Yes, what is it?....Yes, from tomorrow....Yes, I'll send them as quickly as possible....Goodbye!' Then, Benze, who was wearing her pale blue nursing uniform with its white starched apron, walked quickly over to the ship's bell in the dining room, and rang it, furiously. We all ran after her, still holding our ears when she called us to sit down together. What on earth was the matter? 'Quiet!' yelled Benze. The adults, and all of us children, settled down. First of all, Benze explained what measles was. 'It's a very contagious disease, and you get small, red spots on your skin, all over the body. Check carefully to see if you get the spots, and if you do, come and tell me right away. As well as the spots, with measles you develop a high fever, and you can't tolerate daylight. You feel really ill, and sometimes you have to throw up'. It was the Principal Binzer who had called Benze, telling her that because there were so many sick people in the town, it was considered to be a 'measles epidemic'. 'A what?' we asked. But we knew that things were really serious. Benze told us,

> As of tomorrow, there will be no teaching – neither at the Greenlandic, nor the Danish school. The whole school building is being converted into

a "hospital". Everyone who is healthy needs to go to the school and help, because everyone, from babies to old people, can get measles.

Some of us girls hadn't been infected, and we left for school at eight o'clock the next morning. A nurse helped us to set things up. We had to go into a classroom, where there were four mothers with their babies. They were so sick that they couldn't breastfeed or change their babies, so we had to bottle-feed the babies. We were each given diapers, bibs, and feeding bottles. It was almost like caring for living dolls – it was fun, and the babies were so cute. Once the babies had finished their bottles, we changed their diapers, and put them into cots, next to their mothers. The mothers would smile, weakly, and tell us, '*Qujanaq*' ('thanks'). During the breaks, we tried to find out if what people had been saying was true – that there were patients in every single classroom. And yes, it was true. There were old men in one of the classrooms, and old women in another of the classrooms. Occasionally, they called out to us for help, asking us for a glass of water, or to call the nurse. I was very proud to help – I felt like a little nurse, just like Benze. That's where my dream of becoming a nurse when I grew up started. But at other times, we saw people being carried out on stretchers. What a pity it was. On days like that, I felt really sad, and when I would go to bed at night, I couldn't stop crying. I felt such sympathy for the poor people who had lost their loved ones.

There were sounds of coughing and spluttering from every classroom, and the curtains were drawn, too. The doctors and nurses had no free time at all. I don't remember what our boys helped with, but I do know from my older sister's former classmate, Steffen Heilmann, that they wanted to help. The Chair of the Women's Association, Katrine Chemnitz, organised the relief work for measles patients in the town. Amongst other things, small, woollen baby trousers were knitted. As soon as the girls had knitted the trousers as far as the beginning of the legs, the boys – and Steffen was among them – took over the task. It was pretty clear from looking at the finished results that the boys were not very thorough in washing their hands. The legs of the trousers were actually darker in colour, having been made so by the small, dirty fingers of the boys who had finished the knitting. The boys were also sent into the classrooms to empty the toilet buckets.

Then it was my turn. I have to be honest, I really suffered with the measles. My temperature went up to 41 degrees, and my eyes were light sensitive; I was burning up, and I couldn't do anything. It was horrible. One night, I remember Benze gently shaking me, until I woke up properly. I sat up at the end of the bed. I had been having nightmares – it felt like I was like I was drowning, amongst a whole mass of balloons, which kept landing on me. 'Don't worry, it was just a dream', said Benze, trying to comfort me. She got me a glass of water, and put a cold washcloth on my forehead. During the night, I had a sense that Benze had stuck her head around the door, in order to see how I was doing. Although I was ill, I managed to see that her long, thin hair was hanging loose and that she was wearing a long, white nightgown. I thought that it was strange and that she didn't seem to look like herself at all. I think that I was ill for about two weeks, and when I could finally get up, I felt really happy. I could hear the other kids upstairs, and I tried to go upstairs to join them, but after two or three steps, I collapsed – I was still very weak. Benze heard me fall from inside her apartment, and came out, and picked me up. She told me to go back and rest and to take it easy.

A Close Call

It was a lovely summer day in Godthåb – the sun was shining in a cloudless sky. We were all out together, a bunch of girls and boys from the orphanage, who were feeling proud of our new home-made fishing poles. They had all been cut from a piece of heavy wood and sanded so that they felt smooth in one's hands. We had been given some thin twine from the kitchen, and we had been to KGH, in order to buy some big fishing hooks. We knew about a great place in the bay behind Kuløen. It was three mountains away from home, and we'd often seen small fish and sculpin[8] there. Now we were going to give our new rods their inaugural use. We got off to a good start, all the while chatting away to one another. The sea was as shiny as a mirror; we could see the bed, which was of mixed sand and flat rock, below the various rocky outcrops from where we were fishing. The place where the water became really deep and dark was not far away.

My fishing hook sank down to the bottom, and I could see crabs, where I was fishing. 'Do they want to bite onto my hook? And why do they walk sideways?' I thought. They looked funny to me. 'Look at all the little fish!' someone shouted. 'I can see two sculpin – look at how ugly they are!' We ran back and forth between our fishing spots on the high, smooth cliff faces, admiring each other's catches. The likelihood of falling over the edge was greater in some places than in others, so some people could balance along the smooth ledge, whilst others had to find different routes. Little Kristine caught a crab and landed it. 'Can it bite?' We took turns at touching the crab. 'Oh, yes, it can bite!' we cried out, laughing as we quickly pulled our hands away.

Suddenly, we heard a very loud splash, followed by a scream. Little Karl had fallen into the water. We stood there, stunned and staring – he was thrashing about in the deep, cold, green water. 'Come on, Little Karl, swim back in', we screamed. But we had forgotten that he didn't know how to swim. His mouth was open, and he was swallowing a lot of water, as he slowly moved his arms up and down. 'Oh, no – someone do something!' we cried out, in panic. 'He mustn't lose consciousness'. Big Kristine tried to reach him, but she couldn't. Little Karl sank further into the water, and popped up like a heavy log, and suddenly we could see the whites of his eyes. Throwing the strongest fishing pole that we had towards him, Big Kristine shouted out to Little Karl, 'Grab it, and hold on!' The pole landed right next to him, but by now, we could only see the top of his head above the surface. Big Kristine threw the pole out again, and this time, Little

Karl popped up, his eyes staring in horror. 'Grab hold of it, and hold tight! We'll pull you in!' shouted Big Kristine. But by now, Little Karl was too tired, and his lips had turned blue. We started to cry. Big Kristine pulled on her long trousers, and screamed, at the top of her lungs, 'When the fishing pole hits you on your head, hold on!' She moved further out into the water and threw the pole again. Clunk! The pole hit Little Karl's head, and using the very last of his strength, he grabbed onto it. Big Kristine turned him over and shouted out for one of us to hold onto her arm. She got the help she needed, so she could go further out into the water. Little Karl's eyes had started rolling in his head, and as soon as she was close enough to him, Big Kristine grabbed him by the hair and pulled him in. He threw up the seawater. The weather had become colder whilst we were trying to help Big Kristine to save Little Karl. We quickly took off some of our clothes, and gave them to Little Karl and Big Kristine, whose teeth were chattering, and were shivering from the cold. We rubbed their backs and their arms, so that they could get some heat back into their bodies. Phew – it was a close call.

We were all shaken and dumbstruck by the experience, and we went home in silence. We took it in turns to hold Little Karl, trying to keep him warm. Two of us were only wearing underpants, and two of us had no underpants on underneath our trousers, whilst some of the others had no socks or shoes. We finally reached the mountain by the football field, where we spotted Ane Sofie walking home along the road. We shouted out together to her, 'Anso, your little brother was drowning!' She ran over to us and hugged her little brother, who started crying, and then she helped us get him home. We had calmed down a bit by then, and we were starting to feel a bit scared when we thought about what Benze might say. None of us dared to run on ahead – we would face her together when we were telling her what had happened. When we went in through the front door, Benze and Miss Blom, who were dressed in their blue uniforms and white starched aprons, rushed towards us. In chorus, we said, 'Little Karl fell into the water'. Benze looked at us, and then quickly grabbed Little Karl, without saying a word to us. She shouted at Miss Blom to quickly fill the bathtub with warm water, as she wrapped the frozen Little Karl up in a warm blanket. It was a long while before little Karl stopped shaking. He was put to bed, with extra blankets. Not much was said about the incident. Hot cocoa was made for all of us whilst we put on dry clothes. It was a terrifying experience, and Big Kristine became our heroine. We were grateful for her resourcefulness and thanked God in our evening prayers for not letting Little Karl drown. Doctor Rendal came a few days later. Little Karl had a fever and was coughing a lot. It turned out that he had pneumonia.

Was That Your Mother?

We had been down to Ole's Department Store to buy glossy pictures and had agreed to take a different route home – behind the mountains, past the Green-lander's cemetery, and across the Østerbro bridge. There were new barracks in the square behind the cemetery. Construction was going on all over Godthåb in the early 1950s, and we were talking about that, whilst pointing to the red multi-storey building that was being built. 'Look, the new Queen Ingrid's Sanatorium will be finished soon'. Some of us began to run down the street. 'Look, that was my mother going past', one of the girls said, pointing. 'Where?' we asked. 'There!' she said, and ran after her, shouting, 'Mom! Mom!' We saw her mother dash into one of the barracks where the tradesmen lived. We looked at each other, our eyes wide with surprise. We stood there waiting, whilst our orphanage sister ran into the barracks, after her mother. We saw her disappear inside, and after a little while, she rejoined us. Whilst we were going home, she quietly told me that her mother hadn't even spoken to her. She had just been lying there, on one of the tradesmen's beds, with her back turned towards her. Although she had shaken her mother, just to say 'Hello', her mother hadn't even turned around to face her. We felt uncomfortable, and very sad, on her behalf. None of us said anything to the adults. We knew Benze's attitude towards the Danish tradesmen and sailors. Many times over the years, we had witnessed Benze's frustrations when a *kiffak* came in late for work or did not show up at all. On such mornings, her attitudes were very clear – Benze would pace back and forth, screaming at the top of her voice, 'Where is she? Did a ship get in yesterday? Ah, these dirty harbour girls! If she turns up here again, she'll be fired on the spot. You can't trust them – just like some of these Greenlandic *kiffaks*'. Benze would be dashing around; one min-ute, she'd be in the kitchen, cooking something on the stove; the next minute, she'd be in the pantry, cutting slices of bread; and then, sometimes the telephone would ring, too. Some mornings, everything seemed to happen at the same time. 'Well, let's not be harbour girls when we grow up', we'd whisper to each other.

We had helped the *kiffaks* to collect the various coloured wires that were left over from the rock blasting; they were lying all over Godthåb, with all the new construction that was going on. The *kiffaks* liked to curl their hair, especially if there were dances going on in the assembly hall, or if a ship had come in. After washing their hair, and before it was completely dry, they would take a piece of wire and fold it over. Then, they would take locks of their hair and roll them around the wires. Once they had made a roll, they would twist the ends of the wire together tightly. Sometimes, they needed wire-cutters to unfasten the rolls. For the best results, they would sleep with the rollers in. It must have hurt – the tighter they rolled it, the curlier their hair would be. What a vision they were when they went into town in the evenings, with their bright red lipstick, ostrich skirts, and high-heeled shoes. Benze would shake her head at them.

We had a new nursery nurse named Lili, and we really liked her – she was just so sweet. She had short, slightly curly hair, and a beauty spot above her left eyebrow. We girls followed her around everywhere. She was always happy, and she was very easy to talk to. She was also good at reading stories. But she was also popular in the town. Whenever she had free time, she was always going out somewhere. We'd go into her room to see what clothes she was going to wear, and watch her in the bathroom, making herself beautiful. Lili would put on different types of skin cream, put her hair up, and then take out a small can of Vaseline, and put some on her *tissekone*. We looked at each other and asked her why she was doing that. 'So that it's well lubricated', she replied. We didn't understand what she meant by that, and it seemed to us to be just something that grown women did. She also put on nail polish – even on her toenails. She looked really good. She finished up by putting a dab of Chanel No. 5 behind her ears. How did she ever get so glamorous?

One evening, we saw her walking up Børnehjemsvej, hand in hand with Emil Brandt. We hurried over to the corner window of the dining room. They walked on, and Emil tried to kiss her, but Lili struggled, and she looked up at the windows. Immediately, Benze came through the door. 'Oh, no – I hope that she doesn't see them!' we thought, but we pretended that nothing was going on. We sat down, playing with our bright pictures. It was some time before Lili came in. She looked flushed and happy. We giggled and followed her up to her room. 'Would you like a drop of coffee?' Benze called out, up the staircase to Lili's room. 'Yes, thank you', Lili replied as she struggled out of her coat. Inside her room, we whispered to her that we had seen her and Emil. She put her index finger to her lips and said 'Shhh!' We nodded and followed her down the stairs, laughing all the way to Benze's door. Lili's boyfriend was very exciting – a tall, handsome guy.

'We need to get married', Lili told us one day. 'Is that alright with Benze?' we asked her. It was an exciting time. The oldest children were invited to the church, and the wedding reception would be at the orphanage. Lili had a three-quarter length sleeveless wedding dress, with lace trimming and a small v-neck collar. Her veil reached down to her elbows, and it was held in place by a white tiara. Her white bridal bouquet had a single trumpet flower, amongst the smaller flowers, which were set in horseshoe-shape mount, made from green leaves. Emil was dressed in an evening suit, with a small flower in his buttonhole, and a white handkerchief in his pocket. What a beautiful young bridal couple. That was our Lili. Emil sat between Lili and Benze, who wore a nice black dress, with a gold shawl over her shoulders. Naturally, Emil's brother, Povl, was also there. They looked exactly like one another (Figure 19).

When we got back to the orphanage, the big kids had a table for themselves. The boys wore white anoraks and black pants, we girls had our new checked shirts on, and our pleated skirts. The tables were decorated with five-armed brass candlesticks, and two ordinary brass candlesticks in the middle. It was wonderful to be at such a beautiful wedding, for the first time in my life. After they had married, Lili moved out into the town, to live with Emil. They didn't live that far from my childhood home; actually, it was right next to where the twin dwarves had lived, whilst I was still living at home. That was nice, because now we had someone else in the town who we could visit. We could also visit our old *kiffak*, Lene, who had married a Dane, and lived near Queen Ingrid's Sanatorium (Figure 20).

FIGURE 19   Lili and Emil's wedding

The Wooden Raft

There was a small lake on the left-hand side of Vandsøvej. On some of the lovely summer days, we had a visit from a doctor named Esther, whom Benze knew. She lived in the nurses' quarters. She had the idea of helping the boys to make a raft. They hammered and sawed some empty barrels that they had found and tied a bunch of ropes around, and there they had their wooden raft. Everyone was flocking about when it was time to launch the raft on the lake. Oh, how exciting it was! Jens, the janitor, helped them to launch it. Three boys, wearing their rubber boots, jumped onto the wooden raft, and we all clapped excitedly. We had a new toy to play with. Eventually, everyone who wanted a turn had one. Whilst this was happening, Dr Esther was walking around with Benze. But suddenly, Dr Esther screamed; she had plunged one of her legs into a 'crap-hole'. 'Aaargh!' we screamed, and we jumped to our feet. Her boot was filled with pee, crap, and toilet paper. Then, we all laughed – even Dr Esther laughed, in the end. This was in the days before we got the mains sewerage system, and Jens had to dig holes in the mossy soil to empty the toilet buckets into. He dug a new hole each time and then covered them over again. Dr Esther had plunged her leg into one of the newer holes. That was the time that we gave Jens the nickname 'Crappy Jens'.

FIGURE 20 Lene and Arnold's wedding

One day, some of us girls were out picking blackberries. We had wandered over to the far side of the mountain behind the football field. From there, one had a panoramic view of the mountains, Lille Malene, Store Malene, and Hjorte-takken, and the harbour. We were buzzing around, our faces almost buried in the lovely blackberries, thoroughly absorbed in what we were doing. Then, suddenly, we heard a noise, and we raised our heads. 'Hey, what was that?' we asked each other. 'Look at the crazy boys!' They had dragged the raft down to the bay at the ship's harbour, where they had sailed out, and now they were yelling at one another on the wobbling raft. They were having fun. We looked at each other, as we felt a little worried, so we scrambled a little way further down the mountain. We made a plan to shout out to them, in chorus. Cupping our hands to the sides of our mouths, we bellowed, at the top of our lungs, 'What are you doing?' 'Nothing!' they shouted back to us. 'If you're not smart enough to come back, then we're going to run home and report you'. They swung their arms round wildly and shouted, 'If you don't report us, then you can have a turn'. Without even thinking about it, we scrambled down the rest of the mountain to go over to the boys. Fortunately, we also had rubber boots on; we liked to wear them whilst hiking, because there were a lot of puddles and marshy areas around.

The boys brought the raft over to a rocky outcrop and, with two boys and two girls on the raft at a time, we took turns at sailing it. There was no rail of any sort, so we girls clung onto each other. I stared down at the seabed, which looked green, and there were sculpin and Greenland cod swimming around. Oh, how exciting it was. The boys sailed us out to Kuløen, where we scrambled ashore. One of the boys slipped and got his socks wet, because it was low tide, and the

FIGURE 21 Dr Esther

stones were green and slippery. It was nice to get back on solid ground, but
Kuløen was only a very small island. We could hardly wait to sail back. On the
return journey, the boys deliberately rocked the raft. We screamed and told them
to stop it, immediately. When we got back to shore, I jumped off the raft, and
slipped on a wet rock. I landed right on my behind and bruised my back, too. We
had all ended up with a variety of rips and tears in our clothing. It had become
cloudy, and we started to feel the cold when we were helping the boys pull the
raft up onto the beach. We moored it underneath a rocky ledge and roped it to
a big rock. Then, we had to run home over the mountains, as it was almost half
past six. We shared the secret of the raft's hiding place for a long time.

Next to the small lake where our raft had been, there was a huge rock. We
had found a thick board at the barracks, and we put it in the middle of the stone,
in order to make a see-saw. Two of the *kifakks*, Lene and Karen, came out and
played with us. I sat down on one side of the see-saw, Karen sat down on the
other side, and then we called for the others to join us. '*Kiffaks* on one side, and
kids on the other!' Occasionally, the home-made see-saw jerked away from the
rock. One time, I went over to reposition it, but someone sat down on it at the
same time. My index finger was crushed, and immediately, everyone surrounded
me, trying to comfort me. Oh, how it hurt! My finger seemed to change colours
every day, and eventually, I had to go to the hospital and have the nail removed.

Miss Engberg, who joined us after Lili, was very nice, but she didn't stay for long. Her stomach was growing and growing, and we asked her why that was. 'There's a little baby on the way', she said. When we were back in our own rooms, we whispered to each other, 'How is that possible?' We had never seen her with a man, and she lived at the orphanage.

On Leave

We were curious about the noise. It came from Benze, who was clambering her way up the ladder into the attic. We hadn't heard her unlock the door. 'What are you doing?' we asked curiously, and we climbed up after her. 'I'm getting my suitcase', she replied. 'Oh, your suitcase', we said. Then Benze explained that she was going away on leave, to Denmark. 'What does "on leave" mean?' Well, she was going away on a very long break – a holiday of six months. 'Half a year?' we asked aloud. 'I'll explain at dinner', she said, looking around for her suitcase. The light under the rafters was rather dim as gazed around the attic space. 'There's my suitcase!' came the cry, every time one of us happened across one of their own cases. 'And look, here are all of our skis!' It was very exciting to be up in the big attic.

During dinner, Benze struck a piece of cutlery against the glass that she was holding. On hearing this sound, there was an immediate hush. Benze explained that with she was sailing on the next ship to Denmark, where she was going to be on leave for half a year. The children who had not been in the attic with Benze asked her the same questions that we had asked. Benze explained that 'leave' meant a 'leave of absence'. 'What does a "leave of absence" mean?' we asked. Benze told us that because she had not had a holiday since she started working at the orphanage three years ago, she was entitled to six months of holiday. We hardly dared to ask, 'Who is coming to look after us?' Benze said, 'There's a lady arriving on the same ship that I'm departing on. Her name is Miss Ludvigsen'. We were completely silent for amount and then asked, 'What are you going to do for six months?' Benze told us that she would be spending most of her time in southern Jutland, because she had family in Sønderborg, but she was also going on a trip to Jerusalem and Bethlehem. That sounded odd to us – weird. Did she really need to go to the place where Jesus was born? We looked at each other with big, wondering eyes. When we were walking back from the dining room, we suddenly felt saddened by the news. We talked to each other about how strange it would be without Benze. Some of us girls wondered about how she could bear to be away from us for so long, but most of the boys didn't care one way or the other.

The day came when Benze was leaving. Her hat box was on top of her light green suitcase. She was dressed in a beautiful light-green suit, with a brooch and a cream-coloured blouse, and her hair was nicely styled. She was carrying a khaki-coloured lady's bag over one arm, and her khaki-coloured summer coat and scarf were hanging over her other arm. We made eyes at each other when we saw that Benze putting her beautiful cream-coloured mesh gloves on. Benze was being driven down to the dock by one of her friends, who had a ministerial car. Because we wanted to bid her farewell at the ship, we ran the quickest route over the hills, and out onto the narrow road that led down to the harbour. By the time we got there, Benze was already aboard, talking to the captain – she knew him, too, of course.

We were very excited about meeting Miss Ludvigsen, but we were told that she had already been driven up to the orphanage. More and more people were coming down to the harbour. There were always lots of people around when ships arrived or departed   practically all of the passengers' families, friends, and acquaintances would show up. Police Chief Simony's wife, the Governor's wife, and many others were there to bid Benze farewell. We were hardly able to say goodbye; the adults just stood in the way, the men in their walking suits, with their hats at jaunty angles, and the women with their bright red lipstick, and long, red nails. Finally, Benze came over to us and shook our hands. We looked after her in amazement as she walked back up the gangway. Just imagine – she was going to have such a long holiday! We waved and waved, and we ran along-side the ship for the full length of the dock as it sailed away.

Swing and New Times

'Where is she?' we asked, as we returned to the orphanage. We were looking for Miss Ludvigsen, but we were told that she was resting before dinner. 'She's resting before dinner?' we repeated, exchanging incredulous glances. So we went outside to play. When the bell rang for dinner, we ran inside to wash our hands, and we rushed into the dining room. What we saw was something of a surprise – just looking at Miss Ludvigsen made us feel bad. She was a tiny, ugly old lady – even older than Benze – and she was very thin, and wrinkled, and her eyelids drooped halfway over her eyes. On her head, she was wearing a peculiar hat, which looked something like a turban, and was light purple in colour. We groaned. When she opened her mouth to introduce herself, we gasped again – she had such a screechy voice! 'At eight o'clock, you should all be in bed', she said. 'Eight o'clock?' someone replied. 'But the older ones are allowed to stay up until nine'. 'Eight o'clock', Miss Ludvigsen repeated, loudly. 'I'm the manager now, and I'll decide what happens from now on'. A gloomy mood descended on all of us, and by the time we went to bed, we were already talking about how much we missed Benze.

Fairly quickly, the boys started to ridicule Miss Ludvigsen behind her back, and some of the girls joined in with this, too. Some of the boys – as far as I remember, it was mainly Gâba and Barselaj – would wait on the staircase that you couldn't see from the hallway. They'd listen out from there for Miss Ludvigsen's apartment door opening. When she came out, they'd walk along behind her, mimicking her crooked legs and way of walking, and pulling their eyelids down with their thumbs and forefingers. We all giggled. Sometimes, Miss Ludvigsen would stop, and turn around; and then the boys would stop. But as soon as she moved on again, the boys would resume their impersonation. Fortunately, as it turned out, Miss Ludvigsen was only going to be with us for as long as it took for a proper substitute to be appointed. So it wasn't long before Miss Ludvigsen left, and the new substitute arrived. Her name was Mrs Terna Dahl, who was a tall lady, with brown, curly hair. A big surprise for us was that she had brought her young daughter with her. Her name was Runa, whose hair was completely white, but like her mother's, naturally curly. Terna was like a breath of fresh air. When we first spoke to her, we addressed her as 'Mrs Terna Dahl', but she told us, 'No, there's no need – please, just call me "Terna"'. A new era at the orphanage had begun – one with milder winds, and looser reins. It was really quite nice. Our previous bedtimes were restored, and besides that, Terna was gentle, sweet, and kind to us. She was always humming a tune as she went around the place. It seemed like everything had been turned on its head, in a nice way, in these new times (Figure 22).

The boiler room had been repositioned in the basement, so now we had a new workshop space, and we sawed and hammered away. All of the space in the house was used now; we could even play in the attic. When we got home from school, we could sit in the dining room and have a snack and some juice. We had begun to take an interest in music other than children's songs – and now, gramophone records with swing music had arrived. We listened to them over and over, in Ane Sofie's room. Some of the older girls had seen how people danced to swing music. 'I'll show you what to do', said Ane Sofie, whilst the music from the gramophone was blaring out. She would grab an open door by the handle and use the door as a dancing partner. We howled with laughter – no one could have stopped us, and Terna didn't mind the noise. They were glorious times. We practiced the new dances, and once we learnt the main steps and movements, we girls could dance with each other to the swing music.

Initially, we found it a little strange that Runa could come and go from Terna's apartment as she pleased, but we got used to it. We used to think that she was quite spoiled, with all the nice clothes that she wore. And just imagine having such pale skin! But any of our reservations were offset by the fact that Terna was a modern, youthful mistress. I remember that she was very tall for a woman; in some of the photographs that we had seen, she was even taller than Princess Margrethe.

For Christmas, Terna had decided that we'd have our Christmas dinner in the main attic. Because the attic walls were lined with wide, horizontal panelling, it was easy hang up the big, colourful Christmas and winter drawings we'd done, alongside the huge picture of Santa Claus. On Christmas Eve, we found that miniature Santa hats had been put on top of our sodas and that the tables had been beautifully decorated with little candles. It was all so lovely. At every seat, there was a 'sugar top', which looked like big, chalk-white cartridges, and were made of crystallised sugar.[9] Such things had never happened when Benze was in charge. The boys sat at Terna's table, and we girls sat with the *kiffaks* and Miss Blom. It was a wonderful Christmas Eve.

FIGURE 22   Terna Dahl arrives

On Holy Three Kings (Epiphany) evening, after we had eaten, and the tables had been cleared, a story was read to us. As far as I remember, it was 'Uncle Tom's Cabin'. Suddenly, there was a hammering on the hallway door. Both curious, and slightly annoyed, we looked up. 'Who's disturbing us?' Then, three ladies came in, with nylon stockings over their faces, and who were dressed in ugly clothes, scarves, and old hats. They banged their walking sticks on the floor, and it made a real noise. The youngest children cried, and we older ones gathered closely around the adults. 'What's going on? This is far too creepy', we said. Then, the three ladies cackled and walked out through the kitchen door. It turned out that on that night, it was a Greenlandic custom to go in disguise from house to house, so our *kiffaks* had dressed up. But none of us thought that it was any fun.

'*Far til fire*'[10] had arrived at the cinema. We had heard the rumours at school, and we rushed home to tell the adults about this. It wasn't long before we were all allowed to go to the cinema to watch it. Little Per was so cute![11] We really enjoyed our trips to the movies.

'Remember your aprons!'. That was the instruction we heard when we girls were wearing our kitchen uniforms. It annoyed us to have to wear aprons, but sometimes, we'd forget to take them off again. We always had to clean our own bedrooms – so there'd be quite a row if someone forgot to clean up after themselves – but we took turns at doing the bigger household jobs. When we cleaned the bigger floored areas in summer, after sweeping, we had to sprinkle the floors with water, to get them as dust free as possible. During the sweeping, we would go outside and rub the broom bristles against the stairs, in order to get rid of

FIGURE 23 The Workshop

the dust bunnies. In the winter, we'd push the brooms backwards and forwards in the snow to clean the bristles before and after we swept. When washing the floors, which he had to do weekly, used two buckets – one with fresh hot water and the other with brown soap. We had to clean out all the nooks and crannies. Terna would check over our work, and finally, she'd say, 'Holiday!', by which she meant that our work was over for the day.

Our kitchen tasks included consisted of setting and clearing the tables, and drying the dishes and pans after they had been washed. I liked doing the ironing and pressing, up in the attic. We'd collect the sheets, covers, and tablecloths from the ironing press and fold them neatly. There was the pleasant smell of freshly laundered linens, and we could chat whilst we worked.

'Homework!' That instruction always seemed to come when we were playing a nice game with our dolls – me, Bodil and Agnethe had big dolls, and we used to love playing with them. But when we heard that call, we had to stop what we were doing and go into the dining room to do our homework. 'Take everything from your school bags in with you', we'd be reminded. But the very moment that we'd finished, we'd be straight out, and playing again. One day, when the sun was shining, we planned to go hiking and take our dolls along. It was lovely to be outside, playing with our dolls, and enjoying the view over Godthåb. However, I had some trouble that day. I fell over, and knelt on my doll, thereby 'grazing' it. When we got home, one of the adults patched the doll up.

One Saturday, Terna was expecting guests. Her birthday was on United Nations Day – that is, October 24. We had our dinner early, so that a formal evening setting could be arranged – with the fine damask tablecloth, the crystal glasses, and the whole shebang. Of course, we were curious, so we went to meet the guests as they arrived. Terna had arranged for welcoming drinks, and the town's great-and-good were all coming. They looked so smart; the gentlemen were wearing evening suits, and most of the women were wearing stilettos and black dresses. The regional manager, Mr Hesselberg, and the office manager, Mr Andersen, and their wives were there, amongst others. They entered the dining room via the sliding door into the kitchen. We could hear the tinkling sounds, as the guests raised and replaced their crystal glasses. The jovial atmosphere increased over the course of the evening. Meantime, we children were enjoying ourselves, up in our own rooms. We older ones were permitted to stay up until ten o'clock that night. We were reading the new books that we'd borrowed from the library, and we'd been given buns that night. At that point in time, my room was located directly above the dining room. I could hear that things had quietened down a bit, now that the guests were having their coffees, cognacs, and dessert. On such occasions, cigarettes, cigarillos, and cigars were presented to the guests on a silver tray. 'What are they doing now?' we whispered to one another, as we lay there, giggling. We could hear that the tables and chairs were being moved aside, and the gramophone was being set up, so that the dancing could begin. We crept downstairs to the kitchen and pulled the door to the dining room open a little, so that we could peep inside. 'Try to see Mrs Hesselberg, swinging her legs about!' 'Let me see! Let me see!' we whispered. We decided to go over to the door from the utility room, so that we could get a better view. We were just in time to see Mrs Hesselberg, dancing her way back from the large windows in the dining room. She was laughing wildly, holding her black ostrich skirt up almost to her ears, and you could see her very small black panties – it looked like there was only a single black line between her thighs. We gasped in surprise, and at that moment, the adults caught sight of us. 'Could you go up to bed, please?'. We ran back to our rooms and got under the bedclothes as quick as we could. 'Did you see her panties?' we asked each other, giggling with delight.

A Hello to Helene Kristoffersen

'Who wants to come with me to see the new radio shop?' The man who owned it was called Windstedt. He had a son who was very tall and attended one of the senior classes in the Danish school. The new shop was on Skibshavnsvej, near Mr and Mrs Helge Andersen's big house, and it had two large display windows. People would stop in front to look at the radios, of all different sizes, and the gramophones and speakers. It all looked very exciting, but we couldn't afford to go shopping there.

'There's a school party in the gym on Saturday', someone announced. This was something that we were all looking forward to. Some of the older children from the school were going to be performing, and we were going to be wearing our nicest clothes. Recently, we had got new winter dresses. My dress was dark blue, with a yellow and green pattern, and a white collar. The skirt was pleated, so when I spun around in front of a mirror and squinted a little, it looked like a wreath of flowers. It was beautiful. So the ones of us who had pleated skirts would take every opportunity to spin around – in the hallway and the dining room at home, as well as at the party in the gym. So we were excited, but there wasn't much room to move about in the old gym. When all the performers set up at one end of the gym, there was only standing room for us guests, and some of us had to stretch on tiptoes in order to see anything at all. They put on a little play called 'Mr Moppe and his three sons'. Margaret, my afore-mentioned child-hood friend, played the part of one of the sons. It was so funny – just imagine, Margaret playing a boy! The first time that we had seen this particular group of school students performing, Lene, the daughter of Ole, who owned Ole's Department Store, had been with them. On that occasion, they had performed 'Agnete and the Merman',[12] and the accompanying song was really catchy. I can still remember the refrain – 'And they played, and they played, and they played, with the golden dice'.[13]

We always had to be quiet at 6:30 p.m., when the news was broadcast on the radio. This news was also published in printed form, on a piece of A4 paper, and some of the boys worked as deliverers of these, earning 10 øre for each 'newspaper' that was successfully delivered. I can certainly remember Gâba doing this. Every Thursdays, we would listen to a play on the radio. On one occasion, I recall that there was a crime drama called 'Sonja', which was about a criminal who murdered girls. Whenever the next girl was about to be killed, there would be the sound of someone walking in stilettos – click, clack, click, clack. We could hardly sleep that night! On Sundays, I almost always listened to the requests programme on the radio. I always thought up some task that needed to be done in the dining room, so that I could listen to the programme, whilst I was doing it. Suddenly, one day, there was a greeting for me! I listened closely, and yes, it was indeed for me! 'A "hello" to Helene Kristoffersen at the DRK orphanage in Godthåb, and the requested song is, "There are always boats going back, back to Denmark again"'.[14] The message was from my foster parents in Denmark. I was proud and happy and shouted out, 'There's a message for me!' The others came running in. 'What did you say?' they asked. I raised a finger for hush, and I answered, 'There's a message for me'. I thought a lot about the wording of the message. Just imagine being able to visit my foster family in Denmark!

Some of us girls had nannying jobs, and we could earn ourselves about 20 kroner per month. Once, I was the nanny for a family that had the first *sparkstøtting* (a type of motorised sledge) in the town. It was fun riding on it. Every time the Danish housewives had a ladies' get-together, we would look after the children. I had a longer-term nannying position with Mrs Issing Rasmussen. Her coffee table was covered with a beautiful cloth, and like most of the ladies, she had '*Mågestellet*'[15] porcelain – the cups, cake plates, cream jugs, sugar bowls, and the cake stands. In addition to these, she had real silver spoons and silver candlesticks. Yes, it was certainly posh! Later on, Margaret later told that her mother had invited some of the posh Danish ladies over. Margaret's mother was not considered as properly belonging to this refined set of ladies, because Margaret's father was 'only' the captain of a fishing ship. When her mother went into the kitchen, Margaret noticed one of the ladies discreetly turning one of her mother's teaspoons over, and examining it. She whispered to the other ladies, 'It's only "Two Towers".[16] When Margaret's mother returned, the lady quickly put the teaspoon back in place, and the other ladies just smiled at her.

Summer in Egedesminde, 1955

I've been summoned to Benze's office after we've finished eating; Benze made the announcement over dinner. I'm getting more and more worried, and the other children are looking at me in curiosity. On such occasions, our first thought is always, 'What have I done wrong?' Dinner seems to take an age, but eventually, we're all finished. Agathe's going to come with me as far as Benze's door. Of course, Benze has to talk to someone else before me – she has to talk to the *kiffaks* about tomorrow's meals. Finally, she's ready. She's sitting behind her desk, smiling, and I'm standing in front of her. 'Well, Helene, we're coming up to the summer holidays. I've been in touch with the DRK, to see if there's any possibility of you going home to your mother's for the summer. They would like to grant that'. '"Grant" – what does that mean?' I think. But my heart skips a beat with joy. Benze continues,

> I've also found someone to accompany you on the journey. She's a Danish lady, who is travelling to Holsteinsborg. That's the last town before Egedesminde, and as you are so sensible, I'm sure that you'll be able to manage the last part of journey by yourself.

I'm so taken aback by the surprise – and my resultant feelings of jubilation – that I don't even hear all of the practical details. Agnethe, Bodil, and a few of the other girls are waiting anxiously outside Benze's door. 'I'm going home to my Mom for the summer holidays', I tell them. 'When? Where?' they ask me, excitedly.

I can't remember the last few weeks before I left. We had to sail out from the harbour. Shortly after the war, KGH built some wooden trade ships, and I'm sailing on one of the last of the ones that were built for Greenlandic trade – *Tikerâ*, which was built in 1949. When we got aboard, we shown down to our cabin, and after that, we went out on deck, and looked back towards the harbour. Benze was there, as were many of the children from the orphanage – they had come to bid me farewell. Oh, how exciting it is! We wave until we have rounded the coast, and we can't see each other any more. Now I can enjoy the view of the big mountains: Lille Malene, Store Malene, and Hjortetakken. A little later on, I can see the Kookøerne islands, and then, the flattened mountains of the northern part of the country come into view. When it gets colder, I go inside to warm up for a bit.

The lady who is accompanying me explains to me that she knows a family – Mr and Mrs Stærmose – who live in Sukkertoppen, so when we get there, we'll be paying them a visit. The couple are very tall Danes, and they have children, who I play with for a while. Shortly, we'll be sailing again. The journey seems to take a long time. In our cabin, I think about how quickly I will be able to see my mother, after we get to the harbour at Egedesminde. I am lulled to sleep by the sound of the waves striking against the hull of the ship. The lady who has accompanied me this far gets off the ship at Holsteinsborg. I feel a little stressed, but the reunion with my mother and siblings is at the forefront of my mind. At mealtimes, the waiting staff members are very kind to me. Then, an announcement comes, over the ship's speaker system, that we are now approaching Egedesminde. The announcement is made in both Greenlandic and in Danish. I listen carefully to the message in Danish, otherwise I'll be in trouble.

It's after breakfast by the time we actually reach Egedesminde, which is a flat, rectangular town. There are a lot of people on deck, all queuing up to get off the ship as quickly as possible. But I don't want to stand in the queue – I want to be able to see my Mom first. I look out over the harbour, looking intently at all of the people who are standing there. I'm excited, and I have butterflies in my stomach. 'Where is she?' I think, but I can't see her. Gradually, most of the passengers have disembarked and been reunited with their happy families in the town. I remain where I am. The fewer people that are left, the clearer it becomes to me that my mother is not there at all. Now, I'm crying. 'What should I do? My mother's not here! Is this really Egedesminde...?'. A thousand thoughts are flying through my head. But suddenly, a Greenlandic man is shouting over to me, 'Are you Helene Kristoffersen?' I look at him, speechlessly, and nod. He signals to me that I need to get down the gangway, so I pick up my suitcase, and I walk down to him. He has glasses and is wearing a black Alpine hat, and a black-and-red checkered jacket. We shake hands, and he says, 'Hello, my name is Jørgen Søholm, I'm your new father'. I feel completely gutted, thinking to myself that my father can never be replaced. I ask him where my mother is. He points at his head and says, in accented Danish, that she has a headache. I feel very disappointed. I've missed my mother so much, and she hasn't even come to pick me up – and I've not seen her for a year and a half. Jørgen picks up my suitcase, reaches for my hand, and says, 'Come on, let's go home'. As we pass the KGH, some sled dogs come over to us. I feel scared, so I huddle against Jørgen. He laughs at me and then asks me if I'm scared. I nod, so he shoos them with a '*Sshck, peerniarit*' ('Psssst! Move!'). Surprisingly, they move away immediately, and I feel safe again. Finally, we reach a greenhouse with white window frames. 'Here's my house. I built it myself'. His ten sledge dogs start to come towards us, but he stops them. '*Ingigit*' (Sit down!). They whine, and sit down obediently. There's a dog-feeding shed behind the house, with Jørgen Søholm's big, beautiful dog sledge on top.

Inside the hallway, my mother, big sister, and little brother are smiling, and waiting to greet me. They are completely silent, and they look a little embarrassed.

I feel a great sense of relief, and gently smiling back at them, I say, 'Hello'. My little brother, Hans, says, '*Kutaa*' ('Hello'), and Victoria shakes my hand and says, 'Hello'. Jørgen shows me the way to the room upstairs, which I will share with my big sister. We have a bedside table with an alarm clock. My mother and Jørgen's bedroom is next to ours. We go down to the ground floor via a steep staircase. In the kitchen, there is a tiled stove, a kitchen sink, and a dining table with five chairs. There is a beautiful view of Strandvejen from the kitchen window, and one can see the shipyard, the hotel, and all of the other houses. Next to the kitchen is the big living room, which shares the same view as the kitchen, and also has a window which faces to the east. Jørgen has his organ in the sitting room.

We have tea and coffee in the living room. I am offered the armchair, next to Mom. We smile a lot, talk less, but enjoy ourselves immensely. My heart is brimming over with joy. Just imagine – I'm home. After our coffee, we go outside to play. I'm a little nervous of the sledge dogs, but Victo shows me the adorable little puppies inside the dog house. They are really lovely. The mother of the puppies, Arnakasik, walks around us. Victo whispers to her, '*Peerniarit*' ('Move'). She puts her tail between her legs and moves over a little bit, and we can pick up a puppy each, and sit on a rock, petting them. The one that I am holding just about fits in my palms. It whimpers, and it barely dares to look at me – it's so cute that I decide to give it a little kiss. Mom knocks on the kitchen window, and she's shaking her head at me. Victo explains to me, 'Don't kiss dogs, they have germs'. I think, 'That's a shame'. I quickly learn the names of all the sledge dogs. The biggest of them, the one that looks like a giant German Shepherd, is called Bamse; and the one with the black spots is called Milattooq. I spend many hours with the lovely dogs, especially with the little puppies, who follow me around the house. I sit outside with them in the sunshine, and although one of the puppies nips my finger, it doesn't hurt that much.

It feels both strange and wonderful to be at home with my mother for the whole summer holidays. Victoria shows me around the neighbourhood, and at the beginning, she goes with me everywhere. Whenever I ask about something, Victoria gently responds, as best she can, in Danish. Jørgen thinks that it's great that he now has a 'daughter' who can speak fluent Danish – his own Danish is very heavily accented. My mother cannot speak Danish at all. I unpack and put my new rubber boots out in the hallway. Mom is busy baking, because on the Sunday after my arrival, they're going to have a *kaffemik*,[17] to celebrate my visiting Egedesminde for the first time. Jørgen moves the dining room table into the living room and pulls out the table extensions. 'I wonder who's coming?' I think to myself. In the end, my uncle Ignatius Kristoffersen, his wife Inequ, and all of my cousins show up. I am very touched to see my uncle again – he looks so much like my dad. Jørgen's mother Louise, his father Buuju (Bøje), his sister Marianne, and his brother Aksel, along with all of their children, are also coming. I am completely overwhelmed to suddenly have so many family members. There are also other friends of Mom's and Jørgen's. My uncle knew that I was going to be there, of course, but none of the others knew that Magdalene had another daughter, who lived at a Danish orphanage in Godthåb. They don't understand why I'm not allowed to live with my mother in Egedesminde, especially now that Magdalene is married to Jørgen Søholm, who is a respectable police officer.

It's hard to fall asleep. It's strange that it's always light. I'm told that in Egedesminde, the sun does not set at all during the summer. I feel deeply sceptical about this; so one night, when I go to bed, I set the alarm clock to wake me up at the turn of every hour. Every time that the noisy clock wakes me up, I hurry over to look out from the window and find out that it's true – the sun really does stay up all night here. The next day, I feel very tired.

My brand new black rubber boots have their own spot, out in the little hallway. I look at them happily, every time I go into or out of the house. One day, they aren't in their usual place. I look around for them, but they're nowhere to be found. Victo and I go out to play 'Dad, Mom and Kids'. We fill some empty cans with a mixture of soil and water. We use a rocky ledge as a baking tray, and when we turn the cans out, we have some lovely 'chocolate cakes'. When we go inside for lunch, Hans is also going inside, right in front of us. Even though he's hurrying along, I spot my boots. He's borrowed them, and now they're all dirty. I get mad at him, and I scold him. I don't think that he's understand a word of what I've said. He just smiles at me and says, in broken Danish, 'Rubber boots for bad weather'. On the way into the kitchen, I give him an evil stare. After we've eaten I go outside and clean my boots.

One day, I visit Jørgen's mother, father, brother, his brother's wife, and all of their children, all of whom lived under the same roof, in a yellow wooden house, not too far away. *Aanakassak* (Little Grandma) and *Aatakassak* (Little Grandpa), as the family call them, are strange old people. Little Grandpa has charcoal-coloured, curly hair, and Little Grandma has her greying hair tied into a bun at the back of her neck. I'm only eleven years old, but they aren't much taller than me. We communicate with smiles and gestures. Little Grandma makes tea on the stove, then she pours it into a saucer and hands it to me. I do not understand. She makes the same thing for herself, then she blows onto the tea into saucer to cool it, and then takes a sip. So I do the same thing. Then, she grabs a lump of sugar, dips it into the tea, and sucks on it, noisily. She hands me a lump of sugar, and I do the same thing. It is completely still; in the background, we can hear the alarm clock ticking. Occasionally, she says, '*Iliina*' ('Helene'), and smiles at me. How nice it is. Little Grandma likes to sit on her low chair, which is in front of the kitchen window. As I leave, I say, 'Thanks for everything today', and wave to her. During the summer, I go up there quite often. There's such a lovely, peaceful atmosphere. I feel really happy when she makes an eider duck *suaasat*[18] for me. I realised that I had encouraged her to do this on a regular basis, because later on she told her daughter, Marianne, that it was a shame that my visit hadn't coincided with eider season in Egedesminde.

Fighting Over the Dishwashing

As I mentioned earlier, my stepfather was a police officer. My mother occasionally helped out at the hotel. Victoria had learned how to light the stove, and also had the task of collecting coal for it. There was also the job of fetching the water, which was poured into a barrel with a lid, next to the stove. One day, Victo thought that it was high time that I learned how to do these tasks. She put a yoke with a bucket at each end over her shoulders. We walked the few metres over to the water outlet, where the pipe protruded from the ground, after running all the way through a cliff by the roadside. There was a tap at the top of the pipe. Victo filled the two ten-litre buckets almost to the brim, hung them back on the yoke, and balancing carefully, walked back with them. So then it was my turn. The yoke was easy enough to carry when the buckets were empty, but when they were full of water, it felt like my legs were buckling beneath me. They were so heavy! Water splashed out of the buckets with every step that I took. It went inside my rubber boots, and we laughed and laughed – and there wasn't much water left in the buckets by the time that I reached the kitchen, and had to pour what was left into the barrel. Back and forth we went, giggling and joking, and it took half the morning to fetch water.

On another day, we had to fetch the coal. Before Jørgen left for work, he gave Victo a coal token that he had bought in KGH. This was a small, lacquered brass plate, about the same size as a cinema ticket. At the top of the brass plate, there was a small hole, and either 5, 10, 20, 50, or 100 kilos was written on it, in black. The oval-shaped KGH logo – a polar bear sitting on its haunches, with its forepaws outstretched – was inscribed in the middle of the plate. One could get tokens for either the cheaper Greenlandic coal or for the more expensive English coal. If you needed tar oil or petroleum, there were some small, round brass plates that one used as a means of payment, but Jørgen collected those himself. Victo quietly walked up to the service hatch, with the brass coal token. I walked right behind her, even more quietly still, keeping my eyes and ears wide open. She handed the token to the man in the serving hatch, who passed the order to another man, who then shovelled the coal into a sack. They said something to Victo in Greenlandic, whilst pointing towards me. They had probably asked who I was, because Victo replied, '*Iliina*' ('Helene'). Then, they helped Victo get the 10 kilos of coal onto her back. She had to kneel down a few times on our way home, because the bag was so heavy, and we took a break at every rock that was tall enough to rest the bag on. It took a long time to get the coal home.

The third of Victoria's regular jobs was to wash the dishes. Sometimes, I helped her by drying them. One day, Victo threw the washing-up brush at me and said, 'Your turn now!' So I had to wash up, and that day, Victo dried the dishes, with a crafty smile on her face. The next day, after I'd washed the dishes, she said to me again, 'Your turn now!', handing me the dish towel. 'Well', I thought, 'I should probably dry myself off, first'. It was just too much. Feeling angry, I left my position at the kitchen sink and went into the living room. My big sister came after me and threw the dish towel at me. I dodged it and ran back through the living room, out into the hallway, and back into the kitchen. Then, the chase continued, Victo running after me, shouting in broken Danish, 'You have to do the washing up now! You don't wash up at the orphanage'. Then, it finally dawned on me – what she was trying to say to me was that I had to do the dishwashing and drying during the summer, because she had to do it every other day of the year. I replied to her so quickly that she probably didn't catch every word that I said: 'I'm not doing it, just because I don't live at home for the rest of the year. Don't you think I would live at home if I could? Besides that, I do wash up at the orphanage, so I'm not doing the washing up here every day of my summer holidays'. That struck a nerve. Suddenly, Mom was standing at the doorway, trying to mediate between us. She spoke to Victo in Greenlandic for a quite a while and finished up by raising two of her fingers. That gesture meant that we each had to wash up every other day; I could only guess at what the rest of what she said meant. Not being able to speak the same language was difficult. To my great surprise, I was gripped by a sense of homesickness for the orphanage. Jørgen could sense the bad mood when he got home from work. He sat down next to me and tried to convince me that he was my new father. That made it even worse. Heart-broken, I started crying. 'My dad is dead. You can't just get a new dad. My dad can't be replaced'. I threw myself down on the couch, and I roared. Victo looked over at me with real compassion, and she gave me a clean handkerchief. When Hans came in, he looked startled. I felt completely alone in the world, wondering about why I was ever even born. The rest of them whispered together and left me there to lie down before dinner.

I had been told where the orphanage in Egedesminde was. The next day, I went to visit, and it was nice. It was like an exact replica of our orphanage. I went up the basement stairs and said, 'Hello' to the people in the kitchen. 'I'm from the orphanage in Godthåb', I announced, happily. 'Oh! Well, Terna's in there', they said. It was pretty easy to communicate with the people there, because the language used at the orphanage in Egedesminde was Danish, too. It was a real joy to see Terna. She said that she missed us all in Godthåb, because she could trust us, whereas the children here really belonged in orphanages, and they were not as easy to deal with as we had been. I didn't quite understand that, but it was certainly strange to see a fenced-in play area – when I asked about this, I was told that it was because of the loose sledge dogs. After the tea break, I ran home again, in a good mood. Now I'd found some free space in Egedesminde (Figure 24).

One of Jørgen's sisters was called Judithe. I thought that this was an unusual name, but she was very sweet, and she wanted to talk with me, even though she didn't speak Danish very well. She walked upstairs to the bedroom, smiling, and a little while later, she returned with some presents for me. She held them out towards me, in the hollow of her hand. She pointed towards me and said, 'For you'. It was a pendant, with a beige-coloured lady in silhouette, fixed onto an oval piece of amber, and matching earrings. I looked at the beautiful jewellery set, and whilst I was admiring it, she said, several times, '*Tammajuitsussat*'. Victo

FIGURE 24 The DRK orphanage at Egedesminde

tried to explain to me what this meant, which was something along the lines of, 'A memory of me that you must never lose'. I felt very happy and deeply touched.

'We need towels and soap', I said. Victo handed me what we needed. There was no bathroom in the house, so we had to use a public bathroom when we wanted to have a shower. This was a brown wooden house, and everything inside it was made of wood, too. There were planks on the floor, and the water would seep out through the gaps between them. There was a row of shower stalls, and some wooden benches where you could dry yourself, and there were lines of hooks on which you could hang your clothes. It was great to get out for a walk in the fresh air after showering and to feel really clean.

Mom was busy, packing virtually everything we owned for a camping trip. We were going on a short holiday out in the countryside. We sailed off with Jørgen's brother, Aksel, and his family; Aksel had a small motor boat. We rowed out to the motor boat in a dinghy, which we hooked on the back. It looked funny, just like it was jumping along after the motor boat. When we reached a large bay, Jørgen stopped the engine. We rowed ashore, a few people at a time, and then we had to jump out of the rowing boat onto the rocky shore. We found some good sites for the tents, and the adults unpacked, whilst we children were given a *skibskiks* (Greenlandic ship's biscuit) and a soda. Afterwards, we jumped and played around the rocks, before we were told to collect heather for the fire. The Primus stove had been lit, and the kettle had boiled, so we could all have our tea and coffee.

Another family had pitched their tent nearby. They had some boys about our age, and they played with Hans. They were allowed to take their rowing boat out when they wanted to go fishing, or even when they just wanted to paddle around. After watching them for a few days, I got Victo to ask them if I could have a turn. They nodded and told me to sit down on the middle board of the rowing boat. I leaned over to one side so that I could see seabed below the clear, green water. It was strange to be able to see the seabed so clearly; it was almost as if I could reach down and touch it. I spotted a Greenlandic cod and a sculpin.

The two boys said something to me, and I answered them in Danish, telling them that I couldn't speak Greenlandic. They looked at me for a while, and then they started whispering together. I smiled at them. They began to wobble the rowing boat from side to side, gently at first, but then faster and faster. I told them to stop, but they didn't seem to understand me. I was scared that the boat was going to sink; it was pitching so much that a lot of water had come in. 'Stop!' I shouted, and Victo, who had seen what was going on from the shore, rushed to fetch Mom. I had started crying, and Mom started yelling at the boys, and they stopped. Victo wrapped her arms around me, and we went back to the tents and had a cup of tea and a *Mariekik* biscuit.

On our way back to Egedesminde, we stopped off at some islands, in order to gather seagull and tern eggs. I put some eggs into an empty milk box, for Jørgen's mother. She was always very kind to me, and she had grown too old to come along on trips like these. I thought that was a pity, and the day after we got back, I took the eggs over to her. She was really happy.

Jørgen told me, 'On Wednesday, the coastal ship *Jutho* will be arriving. You'll have to sail back to the orphanage on her'. Amongst them, he had the best Danish. My Mom had started washing my clothes in the laundry tub outside; she scrubbed the laundry on the washboard. 'Oh, no, why can't I stay?' I thought. I was happy playing with Victo and the puppies, and we were getting better at understanding one another. In the evening, when we went, to bed, I was feeling very sad. My whole family, and also my new family (through Jørgen) came down to the harbour to wave goodbye. I don't even remember who I travelled back with, or anything about the journey – that's how sad I was.

Holidays in Qooqqut

Benze used her good connections once again, so that we could all go on a summer holiday to Qooqqut. We sailed out from the harbour. It looked beautiful in Qooqqut, and as soon as we arrived there, I realised that I already knew the place. I had been there on a summer holiday with my Mom, Dad, siblings, and my cousin Bibi, when I was four years old. All of us older girls had to sleep on mattresses, in a tiny wooden house. There was no floor space for anything but the mattresses, and the ceiling was so low that we could hardly even stand upright. Every morning, before we helped each other to make up our bed spaces, we had to shake out the blankets that we had slept under. There were loud shrieks from the bathroom each morning, because there was no hot water, only ice-cold water from a well. We poured it out from a watering can, into an enamel wash basin. We said that it was so cold in the bathroom, even the toothbrushes used to shiver!

The manager at Qooqqut was an Icelander. A few prisoners worked there, whilst they were serving out their sentences. They worked out in the fields with the sheep, and also in the garden, where they grew beets, lettuce and radishes. One of the prisoners had his wife and children with him. When we first learned that the men were prisoners, we felt a bit wary, but once we got to know them, they seemed to us to be good men. They didn't say much to us, but they showed their kindness in letting us ride the cute little Icelandic pony that they had there. The sheep grazed on the hillsides, and when we walked there, it was helpful that they had trodden in some paths that we could follow. The paths continued for a long distance. When we came across any sheep, we used to 'baaah' at them.

When one gets towards Qooqqut, there is a very high mountain on the left-hand side, with a broad flank which slopes down from its peak to the water. A few times, we tried to reach the top, but we never succeeded. It was difficult to hear the shout for dinner from the mountain, but there was also a bell as well as the call. When the bell rang, it was great fun to run down the mountain, but in some places, this was difficult, because of the thick vegetation. This was like a small willow forest, and it was fun to play hide-and-seek there. There was also a large grain field, to one side of which was a path, leading from the Icelander's house down to a red building. The field was fenced, with a single gate. We had fun racing one another along each side of the fence, when the grain had been cut. The hay was stored up in a loft, and we could play there, too. One day, when some of us girls were playing up in the hay loft, we suddenly heard a lot of noise. We came down and followed the source of the sound and found that slaughtering benches had been set up and that the lambs that were to be slaughtered were being positioned on top. The boys stood around the benches and eagerly watched the slaughtering process. We girls ran off, crying and screaming, and went back to the hay loft to play with our dolls.

In the red building, there was a kitchen and a dining room, and also a utility room, which we used as a bathroom, and a couple of bedrooms for the adults. On the right-hand side of the house, there was a firewood shed, and there were some sacks of potatoes in the bottom. We were always hungry, and some of the boys used to take some of the potatoes and peel them with their pen-knives. We had some slices of the raw potatoes, and they tasted good.

There was a broad river, which ran along the right-hand side of the field, and out to the sea; at low tide, it was easy to jump between the rocks. There was big, lush bank on the other side. We could run around and play as we pleased there, and all over and around Qooqqut. We picked blackberries and blueberries, and sometimes, we even found cranberries. We ate some of the berries whilst we were picking them – the juice would make our lips and teeth blue. When we were thirsty, we drank from the river, and occasionally, we could see salmon swimming there. The smell of the Greenlandic summer was wonderful. We also freely gathered *qajaasat*, which is a Greenlandic plant with beautifully scented white flowers. There were also masses of purple *niviarsiat*.[19] One day, we had wandered quite a long way on the other side of the river, picking the berries, and we suddenly heard the sound of bells. 'Hey, what's that?' we said, almost in unison. At first, we couldn't see anything, so we carried on picking. Then, we heard the bell sound again, and this time, it seemed closer. We looked up, and suddenly, a herd of reindeer appeared, on the top of the nearest slope. We screamed, jumped up, and raced back towards the river. We'd never crossed the river so quickly before, and more than one of us got their socks wet!

One day, there was a commotion going on at the table. Everyone had sat down for dinner, and as always, there was a count to see who was there. There was more counting than usual going on, and the adults were whispering to one another, looking somewhat perplexed. We children looked around at one other, and we noticed immediately that it was Little Kristine who was missing. 'Who's been with her?' We looked around at one another. 'I have', said Dorthe, in a low voice. 'Where?' asked the adults, speaking as if in a single voice. 'Over at the river', said Dorthe, looking down. 'Oh, no!' said Benze, as she looked at her wristwatch, 'It's high tide! Fetch David, and tell him to bring the horse', she cried out to one of the *kiffaks*. Everyone jumped up, and we ran over towards the river. David quickly overtook us on the horse. Down by the river, we saw the now-sobbing Little Kristine. 'There she is!' we shouted out, waving to her. Then, the adults arrived, and David was told to pick her up, carefully. The tide had risen, and the stepping stones that we used as a crossing place were now covered by the roaring river. David rode the horse out across the river, and when he reached Little Kristine, he gently picked her up, and she wiped her eyes and smiled. She was allowed to ride home for dinner. But before we were invited to start eating, we were given a stern talking-to about making sure that we didn't go off playing on our own.

The older boys always looked happy when were allowed to borrow the rowing boat and go out fishing for cod. We used to watch them row far away, until they looked like little specks out there on the water. Sometimes they caught enough for all of us to have cod for dinner. The prisoners' house was just to the left of the red house where we ate. David's wife was always at home. They had some smaller kids; I don't remember how many. We played with them for a while; they were cute children, and sometimes we went into their house. It was nice in there; we could have a cup of tea, but given that they did not speak any Danish, we obviously couldn't speak with them.

'There's a fire! There's a fire!' we cried out, as the smoke reached us. But it turned out that the Icelander was burning some brambles. He stood at the end of a long mound of earth, which looked like a thick pipe, covered with peat. At the end of the mound there was a barrel, in which a row of salmon was hanging up. He explained that he was smoking salmon. They were going to be stored over the winter, but we were each allowed to taste a little bit. They tasted amazing.

A few days before travelling home, we were all told to go and pick blackberries. We had to pour the blackberries that we had picked into wooden boxes, and we'd bring them home to the orphanage, where we would store them until the autumn. 'Don't eat them all!' said the adults, who were watching this task closely. Amongst other things, the berries were going to be used for blackberry dessert with milk and sugar, blackberry porridge, blackberry jam…. Mmmm, it would be delicious! Everyone helped with the packing on our last day in Qooqqut. It was hard to say goodbye to this lovely holiday paradise. It was also difficult to fit everything onto the ship's deck – it was covered with our luggage, and all of the boxes of blackberries and blueberries that we had picked. The boxes were not sealed, so whenever we children saw our chance, we'd grab a handful of blackberries. However, the adults quickly discovered our trick, and after that, the *kiffaks* kept a close watch on the boxes. After about four hours of sailing, we arrived at the harbour, where a truck was waiting there to take everything home. We children and the *kiffaks* walked back to the orphanage.

A short while after getting back, I woke up in the night. I had an itch in my right armpit, and I scratched it so much that it bled. Later on, my armpit started to hurt a great deal. The area was inflamed, and Benze sent me off to the hospital to have it examined. My armpit was red and swollen and infected in three places. The doctor called the orphanage and told me that I'd have to be admitted to the hospital. It turned out that I had abscesses; I'd caught some infection or another from the prisoners' children in Qooqqut. When I was getting the abscesses removed, I had to lie down on a small surgical table, where they cleaned my armpit. A nurse held me by the right arm, whilst the doctor jabbed me with something he called 'anaesthetic', and told me to count to 100. I talked and I talked and I started to wave my hands around whilst I was counting. So another nurse was called for, and I ended up being held by both of my arms. I heard someone say, 'She needs another shot, she's still conscious'. 'Start counting again'. Then, I had a feeling of complete dizziness, and suddenly, I was away. I woke up to find that some of the orphanage children were messing about next to me. 'Helene! Helene! Wake up, you've had an operation!' they said. I felt completely confused. The nurse had allowed them to wake me up. From that day onwards, the adults were even more particular about us washing our hands.

Tuberculosis

One day in 1955, Benze called me into her office. She had been informed that my mother had been admitted to Queen Ingrid's Sanatorium, which was known as 'Sana'. It gave me quite the start when I heard that my mother was back in Godthåb. 'Mom! Mom!' ran through my head. I didn't understand what had happened, but Benze explained things to me. Queen Ingrid's Sanatorium was inaugurated in 1954, and it was a tuberculosis sanatorium. So my mother had tuberculosis. 'Oh, no!' I thought. That was the terrible disease that my father had died from. 'Children under the age of 14 are not allowed to visit', Benze told me, but she had already talked to the doctor, and he had agreed that I could visit her once in a while, because she lived in Egedesminde, which was a long way from Godthåb. The next day, I went down to Sana, alone. I was upset, and I would have liked one of the adults to come with me, but Benze had said that I was too old for that. It was quite a long walk down to Sana. Along the way, I noticed that a lot of new construction was underway. They were building storehouses next to the Greenlanders' cemetery. It jarred me as I passed, because my father, my uncle and my grandfather were buried there, and I thought it was wrong to build storehouses so close to the cemetery. Finally, I could see Sana, with its low, yellow buildings. I found the main entrance, and fortunately, a Danish nurse came along, and she showed me the way to ward D3. It was a four-bedded room, any my mother was lying there, with her back turned. I walked up to her excitedly and patted her on the shoulder, she woke up. 'Ah – *Iliina?*' ('Helene'). Then, she sat up in bed and gave me her hand. I took a deep breath and shyly looked into her eyes. In this moment of quiet joy, it was hard to keep the tears from falling down my face. She smoothed out the bedclothes and showed me her abdomen; and now I could see that she was pregnant. 'When will the baby be born?' I asked her. '*Juullingajalerpat*', she replied, holding out her hands in puzzlement. 'What does "*juulli*" mean', I wondered. She was still smiling at me as she opened the door to her bedside table, to offer me a cup of coffee. I accepted it and said thank you. She nodded in acknowledgement.

I got a chair to sit on, and even though we couldn't speak to one another, we enjoyed each other's presence. Mom said something to her fellow patients, and occasionally my name was mentioned, so I figured out that she was introducing me. I smiled back, then things got quiet again. After a silent half hour, I went home. When I got there, I hurried up to my room and curled up in a foetal position on my bed. My chest hurt, and I lay there, quietly crying, until it got dark. I wanted to talk to my Mom – to ask her more about her tummy, and to tell her about how much I had missed her. I wanted her to bring me home with her when she was discharged. I wanted to tell her how truly lonely I felt, even though there were lots of children at the orphanage. I wanted to tell her that I often cried myself to sleep. 'Oh, Mom – take me home!' I sobbed inside myself. Then, the bell rang for dinner. Hurriedly, I wiped my eyes and wandered down to the dining room. I was one the last to arrive, and it felt as if everyone was looking at me. Before Benze invited us to start eating, she announced to everyone that my mother had been hospitalised and was at Sana. Agnethe gave my hand a small squeeze under the table, and the others all glanced over at me. I blushed and felt like crying again.

After dinner, the older girls came up to my room with me, in order to hear more about my news. I said that my Mom would have to stay in Sana, but I didn't know for exactly how long. Instinctively, when something happened to one of our family members, we all sympathised with one another. There was always a feeling of support, even though some things were very difficult for us to talk about. We never dared to completely open ourselves up to our peers – and not to the adults at all. There was never the full level of confidentiality, just with one other human being, that we needed. But it was great to feel a stroke across my cheek, and to be told that it was a shame for me.

My mother gave birth to my little sister Rissa on December 7, 1955. I couldn't go down to see her until a week had passed after the birth. When I did get to visit, I wasn't allowed to be in the same room as my sister, because of the risk of infection – so I couldn't see her properly, or touch her. I had to stand in front of a little window, and watch someone else searching for her, because there were lots of other little black-haired babies. Finally, they found her, and they tilted her cot up a little bit, so that I could see her face. She looked really cute, with big brown eyes, and very thick black hair. One day before Christmas, I was informed that my mother and my little sister had gone back home to Egedesminde, on the last ship. When I went to bed that night, I was gripped again by a profound feeling of loneliness; I felt deeply unhappy and absolutely hopeless. I could no longer dream of going home to my mother and my siblings. They had been taken away from me again.

In 1955, a new medical ship was built, for the purposes of studying tuberculosis. The entire population of Greenland was to be examined, and at school, we were informed that this included schoolchildren. We were very excited about this. We walked two by two, in a line, all the way from the school to the harbour, where the new medical ship, *Misigssût*, was berthed. The ship was fully furnished and equipped with modern X-ray equipment; it sailed from town to town, with a doctor and nurse on board. We were told that 'When we get down to the ship, we'll have to have an X-ray taken, and then we'll be jabbed in the arm'. After walking for half an hour, we were close to the harbour, and we could see the ship. Immediately, I thought that it looked like a miniature version of the passenger ship *Umanak*. I thought that it looked cute. When we reached the quay, the tide was low, so the gangway sloped upwards. My stomach churned, and I felt a bit uneasy, when I looked down into the water between the ship and the quayside. When we got on board, we had to wait in a long queue, before taking off our blouses and going into a dark room. We had to stand with our chests pressed right up against a cold plate. Then, we were instructed to breathe in, deeply, and to hold our breath until we heard a 'click' from inside the glass booth, where the nurse was standing, wearing a heavy lead apron. After that, we had to queue up again and roll up our left sleeves, so that we could be jabbed in the arm. 'Ow!' I thought, but it didn't hurt as much as I thought it was going to. The needle that they used was very thin. They had injected a serum, and later on, they would be able to see if there was a reaction or not; a swelling would indicate that we had tuberculosis.

Notes

1 Cape Farewell is the southernmost point of Greenland, known for being navigationally challenging. Prince Christian Sound is a more sheltered waterway due north, which separates mainland Greenland from the Cape Farewell archipelago.
2 This is Tine Bryld's account of what happened to the twenty-two Greenlandic children who were taken to Denmark in 1951. The book's title translates literally into English as, '*In the best sense*', but I have preferred to use '*With the best of intentions*'.
3 The mild expletive '*lorte*' translates as something like 'crappy'; so I have translated the nickname 'Lorte Jens' as 'Crappy Jens'.
4 '*Madam Blå*' [literally, 'Madam Blue'] is a popular brand of enamelled kitchenware in Denmark. The coffee pot is considered a design classic.
5 A traditional Danish folk song (the title of which translates roughly as, 'The Crested Hen'), with lyrics written by Kristen Kastensen in 1850.

6 We have come across this song before – see chapter 4, endnote 28.

7 This is another old Danish song, from the early twentieth century, the title of which translates as 'In a hospital bed'. The lyrics, by Jodle Birge, tell of a heroic sick child who asks to go home for successive religious holidays but is told by the doctor that she is too ill to do so. The girl dies at the end of the song.

8 The European sculpin is a common fish in Arctic and sub-Arctic waters and is found along the coasts of Greenland.

9 A 'sugar top' (in Danish, '*sukker top*') is a conical, crystallised sugar treat. The shape comes from the way in which refined sugar was originally mass-produced in Denmark.

10 See chapter 4, endnote 21.

11 At five years old, '*Lille Per*' ['Little Per'], played by Ole Neumann, was the youngest of the children in the family.

12 'Agnete and the Merman' is a story that dates from at least the early nineteenth century and is known throughout Scandinavia. It tells of a merman who attempts to persuade the titular Agnete to leave her children and to go and live with him under the sea instead.

13 From the refrain, this seems to have been an old Danish song called '*Princessen sad I højeloft*' ['The princess was sitting in the attic'].

14 A recording of this song, '*Der går altid både tilbage*' ['There are always boats going back'], performed by Holger 'Fællessanger' Hansen, was released in 1951.

15 '*Mågestellet*' ['The Seagull'] is the most famous and popular tableware that is produced by the Bing & Grøndahl's porcelain factory. It was created by Fanny Garde in 1895 and had become known as 'Denmark's national tableware' by the 1950s.

16 '*Totårne*' (literally, 'Two Towers') silverware was a more affordable brand than certain others at the time.

17 A *kaffemik* is a traditional Greenlandic social gathering, held to celebrate a range of events. Coffee and cake is always served, sometimes along with a variety of hot foods.

18 *Suaasat* is a traditional Greenlandic meat (seal, whale, reindeer, or seabird) soup or stew. It usually includes potatoes and onions, seasoned with salt, black pepper, and bay, and thickened with rice or barley.

19 *Chamaenerion angustifolium*, which is known in North America as fireweed, and in Britain and Ireland as rosebay willowherb.

6

1957–1960

My Little Brother Hans Dies

My little brother Hans was the only boy in our family and was named after our Dad, Hendrik. Hans Henrik Kristoffersen was born on November 3, 1946. He had lighter coloured skin and hair than the rest of us; his hair, which was thick, just like mine and Victo's, was actually a dark blonde colour, and he had big, light brown eyes. One winter, when he was two years old, he was photographed outside our house, with a little girl and her mother. People said that the little girl was his 'girlfriend'. They also said that a little boy from the Islandsdalen was my 'boyfriend'. I was deeply offended, even though I didn't know what a 'boyfriend' really meant, but the teasing way in which it was said – and the loud laughing that accompanied it – was enough to offend me. It was customary that the person born closest to one's own birthday was called one's girlfriend or boyfriend. When Hans started walking, my mother made a sailor suit for him. Every time he grew out of one of these, she made him a new one – sailor suits were very fashionable for little boys in those days. When he got older, my mother made him a new white anorak every year. In the winter, he wore a hat with earflaps that could be folded down and fastened, and a flap over his forehead, which could be folded down over his eyes, if there was a blizzard. Hans was a cheerful boy, who was always outside playing with his friends. Of the three of us siblings, he was the one who looked most like my Mom, although he was fairer than she was in complexion. 'Hansi!' my mother would cry out, when she was calling for him. It sounded much nicer than pronouncing it the Danish way.

Kunuunnguaq lived in a big red house, down the hill from my grandfather's house. He was one of Hans's classmates. In the summer of 1957, Kunuunnguaq got a brand new bike. Hans often went down there to play with him, and Kunuunnguaq used to let Hans borrow his bike. They both found it difficult to learn

DOI: 10.4324/9781003241843-8

how to ride it, but eventually, the two of them took turns cycling back and forth between their homes, even when it started to snow. One time, though, the surfaces had become very slippery. The slopes at the sides of the roads were covered with large, sharp rocks. My little brother slipped on the bike very badly; he fell off, and rolled down, hitting the sharp rocks, one of which penetrated his stomach. He had four operations, but he got weaker and weaker. I had no clue about any of this, because I was in the orphanage in Godthåb at this time. But on October 20, 1957, I was called into Benze's office. 'What have I done now?' I thought. Benze asked me to sit down in the chair in front of her desk, and she looked very serious. There was a telegram on the desk in front of her. 'A telegram has come for you', she said. Then she read it:

> Hans overturned bicycle STOP four stomach operations STOP died at 6 p.m. STOP on 19.10.57 STOP Magdalene, Jørgen, Victoria STOP.

'Stop, stop, stop', I thought. 'What she's reading can't be true. He's only ten years old, he can't be dead'. I didn't say anything, but the tears began to roll down my cheeks. I just let them fall, and the first thing I thought about was why I hadn't let him borrow my new rubber boots. Time stood still. We got up, and I almost sleepwalked up my room; I threw myself on my bed, and I cried my heart out. Big Kristine, Agnethe, and Bodil came in and asked me what was wrong. It was only when I told them that I realised that my little brother really was dead. They didn't understand me; they said, 'That can't be true, kids don't die. It's only old people who die'. 'Oh, no', I replied, 'Hans is dead, just two weeks before his eleventh birthday. Poor mother, Victo, and Rissa. But Rissa probably won't understand anything about it – she'll only be two years old this December' .

Confirmation

The names were read out: 'The following people have to start their meetings with the priest – Bodil, Ane Sofie, Big Kristine, Agnethe, Regine, Dorthe, and Helene'. 'Meetings with the priest?' I repeated to myself. I remembered some of the older children at the orphanage being confirmed, but I hadn't realised that they had to meet with the priest first. The date of the confirmation was set for April 20, 1958, and straight away, I saw that the other children were smiling at me. That was the day before my birthday, so I would have two days of celebrations. The preparations for confirmation, with our Danish priest, Mads Lidegaard, took place in one of the classrooms at the college. There were already hymn books in the classroom, and the priest brought the New Testament in with him. He had taught us religion previously, and he was really nice, with a wonderful sense of humour. Three other people from the Danish school were being confirmed, at the same time as the seven of us from the orphanage – Jette Linde, Sonja Hansen, and Margaret Mikkelsen. Of these three, we knew Margaret the

FIGURE 25   Hans's burial (ten years) (November 3, 1957)

best, because she often came up to the orphanage to play, and we had been to her house a few times. It was also just great to be an all-girl team. We already knew the Lord's Prayer by heart, but learning the Apostles' Creed would be a bit harder; we'd have to practice that at home. That wouldn't be too difficult for us at the orphanage – there would always be another one of the seven of us to practice with.

Benze said that we could make some suggestions about what we wanted our confirmation dresses to look like. The requirements were that the hem should finish slightly above the ankles and that the dress have a small, round collar, and short sleeves. I designed my dress so that there would be a sewn-in ribbon dividing the top part of the dress from the skirt. When I sketched it out, I put a large bow in the middle of the top part of the dress, with a long ribbon, extending to slightly below the waist, and the skirt section rounding outwards from the waist. We went down to the KGH with our seamstress, Gertrud, in order to choose the textured white cotton fabric to make our dresses. We ran up and down the seemingly endless aisles of KGH's fabrics department. When Gertrud was starting work on my confirmation dress, I had to stand very close, and completely still, whilst she was measuring me. A few days later, our Danish childcare nurse, Erna, came down to the KGH with us, in order to help us to choose our new confirmation shoes. They had to be chalk white in colour, with a heel, but we had a choice as to whether we wanted a strapped design, so that one could see some of one's toes, or a closed design; I chose the latter. We had to have our hip measurements taken, because we also needed suspenders and nylon stockings. I had the strong sense that we were heading into the adult world.

When I had been on holiday in Egedesminde the summer before, I had asked my mother about being allowed to grow my hair long. I was really envious of my big sister's long thick braids. Mom explained to me, as best she could, that she didn't know if I'd be permitted to do that, and shortly after I got back to the orphanage from Egedesminde, we were told to come straight home from school, because the hairdresser would be coming. But maybe things could be different. At a quiet moment, when Benze had gone to get herself 'a couple of drops' of coffee, I summoned up my courage and knocked gently on her door. 'Come in!' came Benze's voice, and she smiled when she saw me. I told her that I didn't want to have my hair cut. Right away. Benze's face took on a serious expression. I told her that Ane Sofie and Big Kristine understood that I wanted long hair, like my big sister's. 'Yes', said Benze, 'But all of the girls here at the orphanage have the same short hair, and that's how it should be.' I summoned up still more courage and told Benze that my mother had told me that I could grow my hair long. There was an awkward silence. 'Well then', said Benze, 'We can use that as an excuse, if the other girls start asking me about growing their hair'. 'Thank you', I replied, turning on my heel. I ran up to my room and threw myself onto my bed. 'I got permission! I got permission!' kept running through my head. When the hairdresser came, they called my name and said that it was my turn next. 'I'm not having my hair cut', I said. 'You have to!' the other children all said. Then, Benze stepped in and explained our agreement. 'Teacher's pet!' said Regine, smiling at me. I ran up to our room.

I followed every inch of my hair's growth with excitement. Just before the confirmation, I had the pleasure of going into Ole's Department Store to buy some white ribbons and black hair-clips, out of my pocket money. Big Kristine and Ane Sofie had bought a white hairpin each, and Margaret had a white bow in her ponytail. When the Danish girls started talking about the guests that were going to join them for their confirmation dinner, I felt sad that I wouldn't be able to bring my mother, stepfather, and siblings. Then, I thought that I might be able to get my Aunt Sofiaaraq and Uncle Hans to come, even though I wouldn't be able to speak with them. After school, I knocked on Benze's apartment door. She wasn't in her living room; instead, she was sitting at the desk in her bedroom. I told her my thoughts about my guest list for the confirmation. 'Yes, but I've already sent the invitations out', she said. 'Everyone whose mother doesn't live in town has to bring the Danes that you're nannying for. So you have to bring

Mr and Mrs Gilberg.' For a moment, I became mute; and then I asked, if it was okay with Mr and Mrs Gilberg, whether there might also be room for my aunt and uncle, because they were the closest (and as far as I knew, the only) family that I had in Godthåb. But I was told that there wasn't room for any more guests, and besides that, my aunt and uncle were Greenlanders. I went up to our room, deeply disappointed.

Agnethe was there, and we stood shoulder to shoulder at the window, looking out over the town. We talked about how we had come home to Greenland, but we didn't feel like we were at home. Agnethe had run away, and back home to her mother, many times. On some of these occasions, her older sister had followed Agnethe back to the orphanage, crying. What was going on in their house? Agnethe was still waiting to hear whether her mother would come to the confirmation. She wanted to be a hairdresser, and her plan was to become an apprentice of the young hairdresser in the town, and to live in town with her oldest sister, who was married to a Danish tradesman. At least Agnethe wouldn't have to stay at the orphanage for a second longer than was necessary. I wondered to myself about whether I would ever be able to escape.

When we woke up in the morning of the confirmation day, we went straight over to the window, in order to see what the weather was like. It was beautifully sunny. I bathed, quickly, and I did not dry my hair completely, because I could braid my hair tighter if it was still slightly wet. With my tongue between my teeth in concentration, I made a very straight parting in my hair and put the tops of my braids behind my ears. The day before, I had ironed my hair ribbons. I put the black hair clips in, so that they clipped my hair in place all the way down from the parting, without so much as a strand hanging loose. Then I could tie the ribbons in. Perfect! I pulled my braids back towards the front, and they reached all the way down to my collarbone.

'Come on, Helene! Benze's driving instructor is here, in his car', the others called out to me. I quickly put on my coat, and carefully walked out to join the others, in my new white shoes. It was lucky that the road was still covered in snow, otherwise our white shoes might have gotten dirty. Benze drove some of us up to our ceremony; Meier had already driven the first of those who were to be confirmed down to the church. Lots of children from the Danish school were in the church, and some of our teachers had also shown up. The people who were being confirmed had to hang up their coats, on the hooks in the porch. The excitement was reaching its peak! We were shown over to some rows of folding chairs, which had been set up next to the choir, and there we sat in our finery, alternately smiling and looking embarrassed, in our beautiful dresses. It was all very holy. The priest signalled us to stand up, in front of the altar, because now it was time to say the Apostolic Creed together. On our way back to our seats, everybody was smiling at us.

After the church service, we put on our coats and went outside, where there was an exchanging of telegrams. When we got home, everyone admired our confirmation dresses. The Danes had arranged a light lunch and coffee after the service, at which we got our confirmation gifts from Benze, which were wristwatches. Mine had a black strap, and it fit perfectly on my wrist. My mother and stepfather had sent me a golden Dagmar Cross,[1] on a gold chain. My foster parents in Denmark had sent me a pearl jewellery set, comprising a necklace and matching bracelet and earrings. When the guests arrived, I got – amongst other things – new underwear in white and pink, nylon stockings, a wallet, and some money. We each spread out our gifts on our beds, which had been dressed especially for today in freshly ironed white covers. Back then, my room was just above Benze's apartment, and I shared it with Agnethe, Big Kristine, and Regine.

We groaned when the guests arrived. We had to shake each of their hands in welcome, and then show them all our rooms and our gifts. Lili and Emil were also invited, and we all flocked around them. It didn't interest us nearly as much that the Principal, Knud Binzer, was there, as were a few members of the Council of Ministers, and their wives – 'We don't even know them that well', I thought. I was seated at the dining table against the wall, with the fixed bench. There was a five-armed brass candlestick at the end of the table, and there were single candlesticks on each of the other tables. Little Kristine had been seated at the far end of our table. Then came Mr Gilberg, with whom I was a nanny;

FIGURE 26 Confirmation – Stine

FIGURE 27 Confirmation – Gâba

then Benze, who was sitting next to me; and then Gâba, who was sitting next to Principal Binzer. Gâba used to help at the Binzers, including bringing in the coal, which is probably why they were sitting next to one another. After dinner, coffee was served in both the dining room and Benze's living room. Those of us who had been confirmed that day had the privilege of being able to go in and out of Benze's living room as we pleased. We were allowed to offer the adults a cigar or cigarette, from the small trays that we carried.

It must have been quite some work at the orphanage, with seven of us being confirmed at once. Perhaps that's why there no pictures were taken on the big day itself – not even a group photo. However, the day after, we were individually photographed separately in our beautiful dresses. We met up in town with a few other people, who had been confirmed they before, and took a walk out to Skibshavnsvej, where we spent some of our confirmation money in the shops. The following Sunday, we were allowed to take Holy Communion for the first time, so we were told that we had to go to church. It felt strange having to get up and kneel at the altar, and I was a bit worried about whether I'd be able to swallow the communion wine. When the priest went to put the communion wafer in my mouth, I took it out of his hand, because I simply had to see anything that was put into my mouth. I could hear some of the others from the orphanage giggling,

FIGURE 28 Confirmation – Óle and Albert

FIGURE 29 Confirmation – Greenlandic guests

FIGURE 30 Confirmation – Danish guests

FIGURE 31 Me, two days after my confirmation

down on the benches. Also, it took quite some time before the wafer melted in my mouth. 'That was a weird thing to eat', I thought to myself.

Hospitalised

In the autumn of 1958, I was called in to Queen Ingrid's Sanatorium, for a talk. At that time, my big sister was in Godthåb, working as a shop assistant. I asked her if she would come down to Sana with me. Recently, I had started coughing a lot. Victo came and picked me up at the orphanage on October 16, and we walked down there, arm-in-arm. The weather was lovely; the first snow of the year had quietly fallen, and there was a nice chill in the air. Whilst we were waiting at the entrance to Sana, I had another attack of coughing. Victo said, '*TB-eqqinnaaq*'. 'What do you mean?' I asked, and she replied, 'That sounds just like TB' (tuberculosis). We were laughing about what she'd said whilst we were waiting by the stairs in front of the entrance to the X-ray department. Then I was called into the office of the chief medical officer, Dr Lang. He shook my hand, and then said, 'Hello, Helene. You were examined on *Misigssût*, and we can see a small black dot on your lungs. That means that you've got tuberculosis.' I sat there listening, in stunned silence. He put the X-ray pictures of my lungs up on a light box and showed me the little black dot on my right lung. 'This means that we're going to have to admit you to Queen Ingrid's Sanatorium today.' Quietly, I asked how long I would be in hospital for. 'You'll probably only have to be here for a few months. I'll call Miss Bengtzen, and her to have your toiletries sent down here. A nurse will be coming along soon, and she'll show you to your bed.' The first thought that I had was that this was the illness that my father had died from.

Victo was allowed to come with me when a nurse accompanied me down the long corridors to ward D3, which was down another long corridor, with storage areas, washrooms, toilets, offices, and a day room. There were several four-person bed bays on each side of the hall. The hallway ended at a garden room, which overlooked the Kookøerne islands and Sanabugten. My bed space was right next to the garden room. 'This is Helene', the nurse announced to the other patients, who nodded over at me. Then I was handed a long undershirt, large underwear, and a knee-length bed shirt. After being shown around the ward, and meeting the nurses, I had to go to bed right away, and I wouldn't be able to get out of bed for the first three months. The first three months! The doctor had told me that I was only going to be hospitalised for a few months. I felt so disappointed. Would I really have to stay in bed for that long?

When the nurse had left, the others started talking to me, in Greenlandic. 'Oh, what's happening now?' I thought. A lady introduced herself as Sofia, and there were two other girls there, named Jensine and Magdalene. I nodded and quietly said, 'Helene', whilst pointing to myself. When I didn't dare to say my name in Greenlandic, they started talking together. In order not to feel any more humiliated than I already did, I wriggled down under the bedclothes, turned my back to the others, and lay there in silence, feeling very upset. I was so embarrassed about not being able to speak Greenlandic. There was not a single Danish patient. The nurses, doctors, and one of the cleaning ladies were Danish. The young nurses were sweet to talk to, but there was an older nurse who thought that it was very odd that I, as a Greenlander, could not speak any Greenlandic. I felt like she was bullying me. One day I thought that she had asked me – with a fake smile, and in very poor Greenlandic – '*Aneerpit?*' ('Have you been out?'). I gasped inwardly, wondering why she was speaking Greenlandic to me. In fact, she had asked me, '*Anarpit?*' ('Have you moved your bowels?'). This was so embarrassing, because she was making a laughing stock of me in front of all of the other patients. I was a Greenlander, who did not understand any Greenlandic at all. I could feel the implied barb. I felt very small indeed.

It was very hard to get used to having the injections – a big syringe, with a big, thick needle, first thing in the morning, and then again in the late afternoon. The injections were made at the top of the buttocks; it hurt, and I was bruised by it. The morning medicine, which I was supposed to take from a small dispensing glass, was also unpleasant. The glass was full of something that looked like chocolate, and on top of that, there was a small green pill, then a red one, and then a small white pill. These would crumble in my mouth, and they were very hard to swallow. Every time I had to pee, or needed to move my bowels, I had to pull on a bell cord, and this made me feel embarrassed. Then a nurse would come along with a large, cold iron pot, with a long handle and a lid. I found this very troubling, and I think that's why my gut stopped moving, and why the staff became so interested in whether I'd had a bowel movement or not. Other than that, I didn't get out of bed at all – or only when I had to bathe, which was once per week. During the hospital rounds, a whole bunch of nurses and doctors would come along. They would stand by my bedside table, ask how I was feeling, and the doctors would talk over my head about whether I had eaten or not. After one such inquiry, they said that as I needed to put on some weight, they'd be giving me a Danish white beer[2] at lunch time. This seemed like fun at first, but eventually, I could hardly stomach it. From time to time, a library trolley would come by. Having to lie in bed all the time was boring, so I used to borrow a good number of books at a time, so that I wouldn't have to get up for something new to read.

It was great when Victo came to visit, and she always brought something for me. Because she was an assistant at a general shop in Godthåb, I might get everything from tinned peaches to weekly magazines. But the best thing of all was that she could speak Danish very well now, and we could finally communicate properly. She was living with our mother's aunt, whose house was near the orphanage. On rare occasions, Benze came to visit me. She always brought some of the children from the orphanage with her; they took it in turns to be the 'lucky ones' who were allowed me to come and visit me. They thought that it was exciting. 'They can't be that smart if they think that', I thought to myself.

When they needed to X-ray me in Sana, I had to go to a different ward. The radiography room there was much larger than the one on the medical ship. First of all, I had to sit down on a small, folding seat, which was connected to a giant machine, which was used to raise or lower the seat. The nurse would then put on a heavy lead apron, go into a little room with a window, and then shout to me, 'Breathe in deeply, and hold it!' Nor was it any fun at all when I had to provide saliva samples. 'How disgusting!' I thought, the first time that I had to do this. I coughed and I coughed to try and bring some saliva up, but most of the time, I couldn't bring up enough. After three months, when I was finally allowed to get up, my peers from ward D3 had to help me down to the ward where they tested the saliva. They grabbed me by the arms and walked me down the long hallway. Then I had to cough and cough, and this time, I managed to produce the correct amount of saliva in the testing tube. This was a strain, because the nurses were not satisfied until the greenish sputum appeared. It was also tricky to walk after being in bed for so long. The funniest thing was when we were running along, with them holding me up under the arms. We had a good laugh about that.

When I was finally allowed to go upstairs, I went along to the stores to borrow some of the hospital's clothes for patients who were allowed to be up and about. A nurse held up a dress in front of me, to get an idea of what size I was. She found a green and white dress that fitted me. I also got some slippers, and the same type of boring underwear that all the patients had to wear. When everything had been sorted out, I had to get ready. I took a shower, and I changed into the clothes that I had been given. I wet my hair thoroughly, tied it into two tight braids, and put a pair of green ribbons into it. Soon afterwards, a nurse came by, and told me that I had to start going to the school, which was in a different department. They accompanied me over there, and I met the teacher and the other children. We were aged between five and sixteen; there were five boys and nine girls. It turned out that the teacher, Josef Josefsen, was a catechist from Kapisillit. He couldn't speak much Danish, and neither could the other children. 'They can't call this a "school"!' I thought. But we learned how to draw and paint objects – for example, bowls of fruit. Using gestures, and a few Danish words, Josef Josefsen explained to me that I had to close one eye and hold a pencil out in front of the eye that was open, in order to get an idea of the relative sizes of the fruits in the bowl. An hour later, we were meant to be reading from Greenlandic textbooks. I refused to do this, as I could neither understand nor read Greenlandic, and I did not want people to laugh at me. So I was allowed to read some Danish books from the room which looked like a small library. There were a good many books there, piled high on the small number of shelves. As time went on, Jensine, Magdalene, and I became closer to one another. They took me along when they went over to the other wards, visiting people that they knew from their hometowns of Egedesminde and Jakobshavn.

Sana was divided by a very long, common corridor, which led on from the main entrance. To the left of the main entrance was the X-ray department and the office, and to the right, there was a list of the different wards and departments – D1,

A1, and so on. Some of the wards were for women and older girls; others were for men and older boys; and there was also a specific ward for younger children. Sana was sectioned off from the new blue-and-white, two-storey townhouses, with flushing toilets, where the doctors and nurses lived. There was an enclave of barracks-like housing on the other side of Sana. These were also staff residences, and they were painted in the same yellow colour as Sana's main buildings. I looked out for anyone who spoke Danish. There was a lady who wore a blue robe, and a white starched apron, who cleaned, amongst other tasks. I asked if I could help her with her work, and she really appreciated that. I washed the floor in the living room, and when I had finished soaping it, I fetched a bucket of fresh, clean water for rinsing. I also rinsed out the floor cloth thoroughly and hanged it up to dry. She was full of admiration, and asked me where I had learned to be so thorough. I told her that my foster mother, Benze, at the orphanage had taught me and that she was very rigorous in her training of us to do practical tasks – her watchword was, 'Imagine if Queen Ingrid was coming to visit'.

Jensine and Magdalene came up to me one day, and asked me if I would attend a funeral with them. They explained to me that someone they knew from another ward had died. We got dressed and walked over, quietly joining the others who were gathering around the coffin. A man said something in Greenlandic, and there were a few hymns; and eventually, they said the Lord's Prayer together, before the coffin was carried out to the funeral car. It was a solemn and depressing atmosphere. One day, Jensine and Magdalene told me that one of their acquaintances would be writing something about spending a year in Sana. So many people had been hospitalised with tuberculosis that the sanatorium was overcrowded, and she had beamed with joy when she was told that she could go home the following day. I was glad to go along with them to visit her on the ward for the last time. However, the following day, when my friends came over to pick me up, they told me that she had died. I didn't understand what had happened until one of the nurses whispered to me, 'We tried to wake her up this morning, but she didn't respond'. They had shaken her by the shoulder, but to their great horror, had discovered that she was dead. It was sad and discomforting. So we went to her funeral instead. In my almost nine months of hospitalisation, I attended nine funerals in total.

When we were having a rest before dinner at the sanatorium, Sofie would sometimes go over to her harp, and quietly play it – those were lovely moments. The harp was black, and I'd watch her movements as she played. I used to wonder about how she moved the harp from place to place, and on her birthday, I was somewhat surprised to see her turn the harp upside-down, and use it as a food tray. She was the oldest in our section of eight patients. She had a lovely smile, but she also kept a close eye on whether we were doing what we were told to do. After a while, I was finding it increasingly difficult to take my medication – I just couldn't swallow it. One day when I was watching the nurse prepare the medicine, I noticed an empty medicine container, which was to going to be thrown out. I asked if I could have it, and I put it at the bottom of the closet by the side of my bed. The next morning, when my breakfast and medicine was set out on my tray, I waited until the nurse had left before taking my medicine. As soon as she was out of sight, I poured the hated medicine into my newly acquired container, which could hold about a litre. I repeated that for quite a while – until Sofie discovered what I was doing. 'She won't say anything about it, because she can't speak Danish', I thought. So I thought that I'd get away with it. However, one day, one of the nurses asked if she could take a look in my closet. I stiffened; Sofie was looking over at the situation that was unfolding, and I sent her a wicked stare. The now-nearly full container was lifted out and handed over to the doctor. He opened it, and gave me a long lecture, asking me whether I imagined that I would get well by messing about with my medication. 'You're only cheating yourself', he said. 'From now on, a nurse will be standing by, and making sure that you actually swallow all of your medication'. After that, the rounds staff left our room and went off down the corridor. I hated Sofie for a long time afterwards.

There were around twenty-four patients on my ward. Everyone was embedded in the humdrum of the day-to-day routine; we experienced our lengthy hospitalisation as an eternity. We younger people were also sick of being expected to sleep after lunch, from noon until 2 p.m. One day, Jensine, who was in the bed to my right, patted me on the shoulder as she got up to walked out to the restrooms. She nodded at Magdalene and pointed towards the toilets. We snuck past the office where the on-duty nurse was sitting, one at a time. Jensine had a small pack of cigarettes and a box of matches in her pocket. We all piled into one of the toilet cubicles, locked the door, and stood up on the toilet lid. Then Jensine lit a cigarette, and we giggled and hugged each other. I coughed, and I only let the smoke go a little way into my mouth before I quickly exhaled. I didn't want to take a drag every time, but it was fun to go along with them, and do something a bit naughty. When the cigarette was finished, we'd sneak back to bed, until it was time to get up. There was tea at 2:30 p.m., and after that we were allowed to be up and about until dinner time. One day during dinner, when we were standing on a toilet lid, we were moving about so much that one of my slippers suddenly broke through and went down into the toilet bowl. We were laughing so loudly about this that the on-duty nurse heard us, and came storming in, and discovered what we were up to. Disgustedly, she fished my slipper out of the toilet bowl; I was given a new pair, and then we were chased back into bed. Over dinner, we had a lot of fun in the living room, telling the other patients about what we'd been up to. But the next day, a large delegation of staff came in, and gave the three of us a lecture. 'What did you think you were doing? You've been admitted here because you have tuberculosis – do you realise how truly dangerous that disease is? Do you want to be in hospital for any longer than necessary? Because if you don't, you need to stop the smoking immediately'.

One evening, we whispered the instructions for the next prank that we had come up with around the entire ward. The plan was for all of us to pull our bell cords simultaneously, at ten o'clock. 'Ding! Ding! Ding!' – the sound came from every bed at once. The night-shift nurses came rushing in and had to run around turning off all the alarms. No one ever found out who had instigated it, and it was great fun.

In December, the nurses found out that I was good at making Christmas stars.[3] Our favourite nurse, 'Ras' as we called her (her name was Miss Rasmussen), was young and really pretty, with curly brown hair set up in a ponytail. She asked me if I would make a whole bunch of Christmas stars for her – she'd give me 25 øre for each one. I was happy to agree to this, because I wanted to supplement my pocket money from the orphanage. The rumour about my skills spread quickly amongst the nurses, and I have never made so many Christmas stars! It was a nice thing to do, and it helped the time to pass more quickly. Jensine also had a birth-day, which was nice, because we had cupcakes and tea in the living room, where there was a painting of Queen Ingrid on the wall, and a red standard lamp at each end of the sofa. It was homely. When someone had a birthday, the table was laid with a white cloth, candles, and a miniature Dannebrog. We either cosied up together on the sofa, or took dining room chairs into the living room.

Hans Hedtoft

A new patient had arrived. I first came across her in the bathroom, where she was explaining something to the nurse, and because she was speaking in perfect Danish, my ears pricked up. When the nurse had left, I said to her, 'Which room are you staying in?' She answered me in Greenlandic, but I steeled myself, and told her that I could not speak Greenlandic. She looked at me, curiously, and then told me that she was in the room right next to the office. 'You must visit me, then', she said. I was so happy that I'd finally met someone who spoke Danish. I went over to visit her straight after breakfast. Her name was Qunerseeq Rosing, and she was a tall, beautiful lady, who was twenty-three years old, and we fell straight into conversation; thereafter, I visited her as often as I could. She even had a transistor radio, so we could listen to the requests programme, and the radio news in Danish. It was so nice to get to know her.

On the radio news, we heard about a new state ship that was being delivered to KGH on December 17, 1958. The passenger ship was called *Hans Hedtoft*; it was built to highest ice class specifications and was further reinforced at the vital points. The rescue equipment was also top notch; there were three thirty-five-person lifeboats, a motor dinghy and four twelve-person inflatables, each equipped with automatic radio transmitters. The ship was double-decked along its entire 87.8 metre length and subdivided into seven waterproof spaces. It was a so-called 'space ship', which meant that the ship retained its buoyancy even if one of the cargo holds became filled with water. On January 7, 1959, the *Hans Hedtoft* embarked on its maiden voyage to Greenland. It sounded exciting, a super-modern ship sailing for Greenland.

The day January 16, 1959 was big. From that day onwards, I would be allowed to go outside for three hours after dinner. It was great to get out in the snow and cold. I breathed in deeply, sucking the fresh air into my sick lungs, although it hurt to do too much of that. The first thing I wanted to do was to go home and visit the others at the orphanage. I was tired when I arrived, but they were pleasantly surprised when I showed up. It was nice to finally be home at the orphanage again. I got a cup of tea and a piece of homemade French bread with jam, and Benze let me come into her living room, and look at all the latest photographs. At my next check-up, I asked if I would be able to go home soon. 'Unfortunately, the prospects aren't that good yet, because the patch on your lung is taking a long time to disappear. We can still see it on the X-rays', they told me. I cried, and wondered if that stupid dot would ever disappear. I couldn't sleep, so I started reading 'Robinson Crusoe'.

One morning when I woke up, it felt like I'd peed in my trousers. However, it felt sticky as well as wet. To my dismay, there was blood in my underwear, and on my bedsheet. I rang the bell, and when the nurse came, I whispered to her, 'I'm bleeding'. She took me out to the bathroom, and into the shower, whilst she collected a sanitary pad, a menstrual belt and clean underwear for me. She asked if I knew what had happened. I knew a little, because Benze had told us about menstruation, and some of the other older girls at the orphanage had already started their periods. Then she took me to a repository and showed me where the sanitary products were, so I could get what I needed.

One other morning, I couldn't face having another injection, so before the nurse came along, I whispered over to Jensine that I was going to hide. I went out into the bathroom and stood absolutely still behind the shower curtain. The nurse came into the ward, with the syringe at the ready. Jensine told me afterwards that the nurse had looked surprised – she was calling my name up and down the entire ward, and asking the other nurses if they had seen me. Then I heard her coming into the bathroom. Suddenly, I heard her say, 'Ah, there you are!' – she had noticed my legs below the shower curtain. She lifted up my night shirt, and plunged the horrible needle into me. It felt as if she had hit my hip bone. I was so sore, and I had so many bruises, at the tops of both my buttocks, and I cried. I hated being hospitalised.

I was often went over to visit Qunerseeq, and we got into the habit of listening to the Danish radio news together. I will never forget the first serious news item that I heard, which was on January 30, 1959. On the radio news, they said that the *Hans Hedtoft* had sent out an SOS call at 1:56 p.m. 'Shhh!', said Qunerseeq, although neither of us was saying anything. We followed the news, broadcast after broadcast. The *Hans Hedtoft* had collided with an iceberg. They then said a lot of things that we didn't understand about the ship's position, which was at 59.5 degrees north and 43 degrees west. We heard the terrible news that the ship was sinking and was about to fill with water. It was awkward, as we couldn't think or talk about anything else. We had been told that the ship could not sink, so the whole thing was incomprehensible to us. When we heard that three other ships were on their way to *Hans Hedtoft*'s position, our hopes raised a little. Minute by minute, and hour by hour, the news reports came in. The ship had set sail from Julianehåb on January 29, 1959, at 9:15 p.m. A search headquarters had been established in Grønnedal Immediately after the *Hans Hedtoft* had sent out its first emergency call. Following the news was tense; I'd sailed between Green-land and Denmark myself, on the *Umanak* and *Disko*. I knew how terrible the weather could be at Cape Farewell – giant waves could be washing all over the ship. Even though it was summer when I'd sailed to Denmark, I'd been told that it was stormy at Cape Farewell. And this was winter; even though we had bright sunshine in Godthåb, there were terrible snowstorms down there.

I went out for a walk after hearing about the *Hans Hedtoft*'s distress signal. I was on my way up to Ole's Department Store, and just before I came out onto Skibshavnsvej, I met my friend and classmate, Helga. She looked sad, and when we made eye contact, I could see that she was crying. I asked her what the matter was. 'Haven't you heard about the *Hans Hedtoft*?' she asked me. 'Yes, of course', I replied. With the tears now rolling down her cheeks, she told me that her father had been on board. 'Oh, no!' I thought. Her father was a member of parliament, named Augo Lynge. I didn't know what to say, so I squeezed her hand, and we parted company.

There were news broadcasts about the *Hans Hedtoft* every day. We continued to live in hope that the ship would be found – it seemed unlikely that a brand new, unsinkable ship could disappear without trace, and there were several places in the ship where it was impossible for water to get in. 'The passengers and crew must be waiting to be rescued', I thought. But on February 6, we heard on the radio news that the search was over. Slowly and deliberately, the names of all those who had lost their lives were read out. It all felt unreal, and it made a huge impression on me. We sat there, wordlessly, just staring into space. Qunerseeq had known many of the people who died; I knew Helga's father, and a couple of people from Egedesminde, who my mother and stepfather also knew. A man called Halfdan Egeskjold and his wife were also mentioned; they had been on their way home to Denmark to enjoy their retirement.

In the school at Sana, the students were rehearsing a piece of theatre, which they wanted to perform at various festivals. Of course, it was in Greenlandic; they told me that I should join in, but I definitely didn't want to. They laughed, and yelled at me to get on stage. I kept on refusing, and when they continued to pressure me, I couldn't take it anymore. I started crying – 'Just leave me alone! I can't speak Greenlandic!', and in the end, they let me be. One of the young male patients played the guitar, and the two youngest boys sang. Most of the performers wore masks that they had made themselves; one dressed up as a nurse, and our teacher Joseph wore a ladies' robe and played the violin. It felt very festive, but I missed out on a great deal of what was being said, because I knew only the Danish language. One day, Magdalene had said '*Pilatsissaaq*', when she was talking about a girl whom she knew. I thought that I had understood what she was saying, and I asked in reply, 'How is she feeling about going home to Egedesminde?' At first they glared at me, and then they started laughing. I felt very stupid. Magdalene had been saying that the girl whom she knew was going to have an operation, and I thought that she had already recovered, and had a ticket for the ship home.

For my birthday, there were tea and buns in the living room, and some adults came in from another ward with a gift for me. It turned out that they had put their contributions together to buy a gift that would be presented to me by the chairman of the patient association. It was a red photo album, with a picture of Sana already pasted inside.

One day, when Benze was visiting, she told me that the chief physician had given his permission for me to attend a very distinguished ceremony. I was away from Sana for the entire day on May 25, 1959, so that I could be at the orphanage. Like all of the other children, I had to dress up smartly for the event. Benze was being awarded the Florence Nightingale Medal in recognition of her work at the orphanage. The boys were wearing their white anoraks, we girls were wearing our confirmation dresses and white, high-heeled shoes, and we were taken over to the Council Chambers, where many people had gathered. Ras and some of the other nurses were there, dressed in their nurse's uniforms, with their starched nurse's caps fastened into their hair with clips. When Benze arrived, the nurses formed lines of welcome in front of the main entrance, and after the speeches were held inside the National Council Hall, Benze received her medal, which was fastened onto her green DRK uniform. Finally, Benze gave a speech of thanks. In retrospect, I mainly used my period of 'leave' from Sana to hang around with the other patients (Figure 32).

Finally, during his round one day in July, the doctor told me that I'd be discharged the following day. But being that I'd stopped taking some of the medication for a while, they didn't know exactly how far behind I was with some of it, so I'd have to bring a long prescription list home with me. 'As a bonus', I thought; but there was nothing I could do about it. So finally, I could go home. Some of the children had moved on. A girl named Julie had moved in, but Ole was with the school ship, Eli had started work as a painter in Juli-anehåb, and Albert was painting in Godthåb. Ane Sofie and Bodil had gone away to a boarding school, and Barselaj had started at a high school. Many new younger residents had arrived. I moved into an upstairs twin room with Johanne. It was a little strange to be back again. When I had to take my medication, the other children would crowd around me, looking at the strange things that I was about to swallow. 'Oh, that looks like chocolate', they said. By that time, I didn't find it too dreadful to have to swallow the medicine, even with so many onlookers.

Summer Holidays in Denmark

My last year at the orphanage was 1960, and in the spring, we were given the news that Benze had arranged for some of us to go on a summer holiday to Denmark. Those who would be going were the children who had been in the first team of 'Benze's children', and who were still living at the orphanage – so

FIGURE 32 Benze receives the Florence Nightingale Medal

those who had come to Godthåb in 1952, when we were about seven years old. In collaboration with the DUI department[4] in Sønderborg, Benze had arranged the trip for us who were now the 'big' children at the orphanage – Gâba, Søren, Regine, Sâmo, Little Kristine, Little Karl, Dorthe, Karl-Otto, and me. She was going to make a final effort to give her first flock of children a great experience (Figures 33 and 34).

We were enormously excited and were really looking forward to setting off. Shortly before we went to Denmark, there were official visits to various ministers in Greenland, who were going to travel with us. They had already visited us at the orphanage, so it felt like we already knew each other a little bit. We were travelling on a 'Catalina', which was a seaplane. First of all, we flew to Frederikshåb, where we landed on the water. However, there was no time to stop there, which really annoyed Gâba, because that was his hometown. He was really feeling it, because we were all asking him if we could see his house, and to

FIGURE 33 Goodbye, Benze! With (left to right) me, Dorthe, Hans, and Agnete

point out the various places, but it was difficult to see the town from our landing spot on the water. The next place we got to was Narsarsuaq, where the ministers bought duty-free goods; and we were spoiled by the adults, who bought lots of sweets for us there. After Narsarsuaq, we landed in Iceland. We could spend a fair amount of time there, so the ministers took us out in small groups in taxis, so that we could see Reykjavik. That was exciting. The next stop was Scotland, but along the way, because we had eaten so many sweets, some of us got airsick. We didn't see too much of Glasgow – actually, only the airport – before we flew onto Kastrup Airport.[5] Flying to Denmark was great, because when we first went there, it had taken us fourteen days to make the sailing. We were tired when we arrived at Kastrup, and we were all wondering about who was going to pick us up. I remember that it was a really nice man with dark, curly hair, who drove us in a minivan through the rising heat towards Sønderborg. It was warm, even though we opened the windows, and the van made a strange hissing noise as we

FIGURE 34   Thanks for everything, Benze!

passed other cars, so we weren't able to take a nap whilst on the long drive. People stared at us on our occasional meal breaks – a bunch of dark-haired Greenlandic children was not a common sight at that time.

At last, we reached Sønderborg, where our host families were there to greet us
at the railway station. There was a big rush. I was going to be staying with a married couple – I thought they were very old, and I was disappointed to find out that
they had no children. They lived on Oehlenschlägersvej, on either the second or
the third floor. They had a tiny kitchen, and in the living room, the guest bed
where I would sleep was folded up behind the dining table. Every morning, I'd
have to fold the bed back up and put it away. They didn't have a bathroom – only
a toilet – so we washed ourselves at the kitchen sink. There weren't that many
secrets. My eyes were certainly rolling, because the woman in the house shaved
herself each morning. And so did the man, of course! However, they were not
very talkative, so I was glad that we had a full programme each day, and I could

be with the others from the orphanage, their host families, and the others from the DUI in Sønderborg.

On one of the first days there, we were invited to visit Benze, in her apartment. I was very much looking forward to it, but when I saw her, I was a little disappointed with her appearance. She had become very much older in looks; she was wearing a long sand-coloured coat and had a small scarf tied around her head. But it was nice to be able to speak with her again. She had bought herself a first-floor apartment on Bygtoften, which had a small balcony terrace. It was nice to see her paintings and furniture again, which we recognised from her time at the orphanage. We whispered about that to one another, and also about Benze's breasts bouncing up and down as they used to, whenever she was having fun.

I wasn't in the best of moods, because I didn't really connect with my host family. The other girls saw me crying about this, and a different host family was found for me. So I joined the Bech Andersen family in Algade, who had a daughter, Birthe, who was almost the same age as me. There was also a big sister, who was married, and lived in the apartment upstairs, a little sister, and five-year-old Gert, who was their young niece. The father was a prison officer, and the mother, whose name was Irene, was a homemaker. They had a large, cosy apartment with a bathroom and shared an outside toilet in the yard. They also had a big car, which was a Vauxhall. They were really good people, and I went along with them to visit their family in Funen and to a campsite in Blommenslyst. Birthe and I hung around watching the boys at the campsite. We giggled when a young German boy had the hots for me and called me '*Die schöne braune Augen*' ('the beautiful brown eyes'). One day, we went on a trip to the beach at Rømø. I was surprised to find out that one could drive a car on the beach. When we went back to Sønderborg on another occasion, we visited the family. My 'holiday father' picked a pretty little blue flower, and asked me if I knew what it was called. I didn't know. 'It's called a "forget-me-not"', he said, handing it to me.

In their apartment, they had a bay window, where there was space for a small table. I often sat there with Birthe, watching those passing by – both those in cars, and those who were walking past – drinking coffee, and talking about anything and everything. Occasionally, we'd take a trip on the 'liquor boat'[6] back and forth between Denmark and Germany and have lots of cheap sweets and soda. How different the Bech Andersens were to my first host family! In their home, I felt like a part of the family, and I was very spoiled. We also went around to visit some of the other orphanage children in their vacation homes. It was great to see them, and also to meet the people with whom they were staying. We all mixed really well, and we had a lot of fun – the host families were very hospitable. When we visited the various hosts' homes, we particularly enjoyed their beautiful gardens. We absolutely loved the flowers, and strawberries with cream were the big hit of the summer on the south Jutland dining table. They taught us how to say, '*Venné ge tåch?*'[7] ('When is the train leaving?') in the south Jutland dialect, and we howled with laughter. Little Kristine's birthday was celebrated with layer cake – and many other cakes – and with daisies on the table. We girls wore our best summer dresses, and some of the boys wore their white anoraks.

We usually travelled by bus when we were travelling around Denmark with the DUI people. We'd be in a great mood, singing, '*We have a driver, a driver, he drives a wheelbarrow*',[8] inventing new verses, and shrieking with laughter at each new one. There was a great deal of interest in us at the DUI camp in Aalborg, where we were seen as the 'cute Greenlandic children', and Søren was interviewed on national radio. It was hard to say goodbye to our holiday families, and to Benze, when we were getting on the train to leave Sønderborg. Would we ever see them again? Our next stop was Odense, where the DUI people from Højstrup met us, wearing their uniforms. I quickly felt at home with my next host family, though. As far as I remember, they lived in a townhouse. At the dinner table, one thing that I noticed was that the ratio between meat and potatoes was different. In Denmark, you ate one, or at the most, two pieces of meat, but all the potatoes you could eat. In Greenland, it was precisely the other way around.

The people who had arranged our stay in Fyn had been in contact beforehand with Fyns Stiftstidende,[9] so we were featured in that newspaper several times. On August 11, 1960, the heading read, 'The Greenlandic children met their neighbourhood hosts in Højstrup with happy smiles on their faces. They have no doubt that they'll have some exciting days in Odense'. Underneath one of the pictures of us, the text read,

> Today, the Greenlandic children who are visiting DUI in Højstrup were at the Funen Village,[10] along with their host families. It was a great experience for the children, who studied with great zeal the very distinctive way in which houses have been built in Denmark over the years.

On Saturday, August 13, underneath the headline, '*Guests from the northernmost part of the country*', the text read (Figure 35):

> Currently, nine Greenlandic children from an orphanage in Godthåb are staying with DUI members in Odense, where they have had some eventful days. They have, of course, been to both Hans Christian Andersen's House, and the Funen Village. Here, the guests from Godthaab are photographed with Møllen in the background, and in the second picture, Ulla Nørgård is showing her friendship towards the guests, placing a loving arm around Gâba Schmidt. DUI's Højstrup department is taking care of the children. On Monday, they are travelling to Copenhagen, and on August 19th, they will be travelling back to Greenland. Various companies in Odense have contributed to the children's travel and accommodation expenses.

We sailed home to Greenland, filled with more experiences than ever before. Before we went to Denmark, we knew nothing about DUI, but we will never

FIGURE 35 Newspaper photograph from holidays in Denmark (August 1960)

forget them – and we are forever grateful for all of the wonderful experiences that we had during that summer holiday in the summer of 1960.

Notes

1 These are crucifix pendants, the original of which is an eleventh- or twelfth-century piece said to have belonged to Dagmar of Bohemia, who was a Queen of Denmark. Especially since 1863, when Alexandra of Denmark was given a copy of the Dagmar Cross by Frederick VII of Denmark on her engagement to Prince Albert Edward of Wales (later, King Edward VII of England), a Dagmar Cross has been a traditional confirmation gift for Danish girls.

2 Danish *hvidtøl* (literally, 'white beer') is a traditional beer, which is sweet and malty. Despite the similarities in the English-language translations, at just two per cent alcohol by volume, it is not to be confused with German and Belgian 'white beers', which are usually considerably stronger.

3 *Julestjerner* (literally, 'Christmas stars') are another form of traditional Scandinavian Christmas decorations made by interweaving coloured paper, much like *Juleherter* (see chapter 4, endnote 17).

4 DUI ('*De Unge Idraet*' – roughly, 'Youth Sports') was founded in 1905 as the youth section of the Danish Social Democrats (although in reality, the party affiliation has usually been quite loose). DUI has organised camps, leisure activities and international student exchanges.

5 Formally designated Copenhagen Airport, this is Denmark's main international airport, located in the municipality of Kastrup.

6 Alcohol has been very heavily taxed in the Nordic countries for decades past. Before Denmark, Sweden and Finland joined the European Union, shopping outlets on the

passenger ferries between those countries, and between those countries and other European destinations, had duty-free sections. These provided a rare (and welcome) opportunity for Nordic people to buy their alcohol relatively cheaply. Hence the term '*spritbåd*' (roughly translated, 'spirit boat' or 'liquor boat').

7   This is a strong dialect; the standard Danish is '*Hvornår går toget?*'.

8   A funny military song, often sung by children on journeys (comparable to 'Ninety-nine bottles of beer on the wall'). Most of the humour comes in the refrain, where it's revealed that none of the people who are mentioned can see properly, despite this being necessary to do their jobs.

9   This is a Danish daily newspaper, headquartered in the city of Odense, which serves primarily the island of Fyn.

10   An ever-popular local attraction, the Funen Village is an open-air museum, which was founded in 1942, and features many older buildings, ornamental gardens and old Danish livestock breeds.

7
WHAT HAPPENED AFTERWARDS

How did I manage it, then? I was one of the ones who got through. I worked really hard at school, and I made every effort to do all of the practical things at home in the orphanage. I was a much-sought-after nanny with the Danish families. You couldn't lay a finger on me! I did my homework beforehand, and I got the first of my prizes for diligence in the fifth grade. In August 1960, I was admitted to the Greenlandic boarding school in Godthåb. It was a great day of liberation. Just think, I was going to be studying at the Ilinniarfissuaq College, which my father had attended, too. I lived there with many other young Greenlanders, from all up and down the west coast. The biggest thing of all was that I was no longer a Danish orphanage child. A new era in my life could finally begin. Finally, I could smile again.

One hundred and ten of us students lived at the college. Ane Sofie and Bodil had just completed their second year, and Gâba, Søren, and I were enrolled in the first year. Previously, our study programme had taken two years, but in 1960, an attempt was made to get our cohort to complete the curriculum in just one year – which we did. Those of us who had come from the orphanage tried to sign up for classes in the Greenlandic language. Our teacher, Abel Kristiansen, looked on the class roll, but our names were not there. Disappointed, we went to Principal Binzer, in search of an explanation. But there were no resources to teach us Greenlandic from scratch. So when the others had Greenlandic, I tried to use this spare time wisely. Finally, I would be able to get to know my aunt. She lived near to the college, just behind the church, with her husband and her in-laws. Her mother-in-law gave me a cup of tea and a slice of Greenlandic cake, and my aunt said, '*Iliina*' ('Helene'). Aunt Sofie and I sat there, smiling and nodding gently at one another. I met my little cousins, and when I was there after school hours, I got to know my older cousins, too. The new sense of family togetherness

DOI: 10.4324/9781003241843-9

was strong, and it felt overwhelming – sometimes, when I got into bed, I would quietly cry with joy. When my cousin Maliina started at the high school a few years later, we got even closer, and that went for my cousins Arnaruniaq and Hans, too. I had a family now.

At the graduation ceremony, I was sitting with my classmates. Suddenly, whilst we were sitting there whispering to one another, I found myself being nudged, from several directions at once. My name had been mentioned. I had obtained my diploma, and I was already pleased to have been awarded an excellent grade in Danish, so I couldn't understand why I was being called upon again. Somewhat reluctantly, I went up to Principal Binzer; he smiled happily at me, handed me a gift, and then told the entire assembly that as I had been such a conscientious student and good classmate, the Teachers' Council had decided that I should receive a prize for diligence. I felt very embarrassed. The prize was a book on cookery for the electric kitchen. On the first page, the following had been written:

Helene Kristoffersen, Class 1A
Awarded:
For diligence and good behaviour
COLLEGE AND HIGH SCHOOL
During the school year of 1960–61
GODTHÅB-GREENLAND
Godthåb, d. 24.6.61
K.Binzer

I wanted to continue studying at the new teacher training college, because my then-boyfriend, Hans A. Lynge, was there. I attended the preparatory classes, but when we were hired early, as unpaid substitute teachers for the difficult sixth and seventh grade classes, I asked if I could train as a kindergarten teacher. It so happened that I got an internship at Save the Children's kindergarten in Egedesminde, so now I, as a twenty-two year-old, could finally go home to live with my mother, my big sister and her children Hans and Harriet, and my little siblings Rissa and Asta. It was wonderful to be in the bosom of my family for ten months – being so close to them was an incredible feeling. I needed to work hard, and the children enjoyed the fact that I baked so often. It was also good to see my stepfather's sledge dogs again, many of whom I'd gotten to know when I was on a summer holiday in Egedesminde, and some of whom I'd named. They followed me to work every day and sat waiting for me until I finished. So I'd walk home with a pack of dogs at my heels.

I did everything I could to learn Greenlandic. I listened to as many of the bilingual radio news programmes as I could. I would write down the Greenlandic words that I didn't understand, and then wait for the translation in the Danish version of the broadcast. I was always asking my siblings, 'What does *this* mean?' I was happy whenever I learned a new phrase, but I didn't yet dare to speak the

FIGURE 36 Mum and my two new sisters

language in company, and I found it was humiliating not being able to answer when people spoke to me in Greenlandic. They looked confused, and I often saw judgemental glances. Gradually, I could understand most of the conversations that were going on, but there were almost always a few words that I didn't know.

I thought that I would be able to go straight into kindergarten training after the summer. However, there was some bad news from the Ministry of Green-land, which had a hand in dealing with the young Greenlanders who were going into education in Denmark. I was told that the time I had spent as a trainee in Egedesminde didn't count in Denmark. This was typical: 'This rule does not apply to Greenland and the Faroe Islands'. So I had to do another period of pre-training experience in the spring of 1967, which was six months in Skov-lunde. My advisor found a room for me in Herlev,[1] and I got a bike so that I could cycle to Herlev Station. But making the train connections wasn't easy. I used to stand there thinking, 'Which train should I be on?' One time, I was just about to board a train, when I felt someone tugging me from behind – it was one of the teachers, who had seen me getting on the wrong one. Another time, I asked the conductor to tell me when we reached the station where I needed to change trains. And another time, I fell asleep on the train, and woke up to find that we were at the terminus – it took hours for me to finally get home. I knew that I couldn't continue this way.

Then, the rules for admission to kindergarten training changed again. Now, you had to have different set of things in place before you could start, including a period of working in a house in Denmark, and some higher education. My advisor at the Ministry of Greenland thought that it would be best for me to undertake my higher education studies first, and then she would help me find a place where I could spend the minimum period of four months of working in a Danish house. I felt deeply indignant – did I really have to become a *kiffak*? Unabashedly, my advisor asked me whether there was a specific college that I was interested in attending. We looked at the map, and we chose the one at Ry, as it was closest one to Silkeborg, where my friend Kaaraaraq was going to start training as a nurse. My scepticism about the college stay turned out to be completely unfounded. The place was beautiful, and on the first Sunday, we went out together for walk, in order to get to know the area. As we could bring a guest, I invited Kaaraaraq, and we walked and talked about life, whilst enjoying the Danish forest in autumn. One of the taller guys started to walk very closely behind us, and we couldn't speak in peace – we started whispering together in Greenlandic, wondering whether he was flirting. Eventually, we ran off ahead, but suddenly, he caught us up, and hugged me from behind. 'Stop that!' I said, tugging myself free. Then, I saw that he was a good-looking guy, with a terrific smile, and he had gorgeous blue eyes – the same colour as the sweater he was wearing. And that was Ove.

Over the first days, because so many people had signed up for these subjects, we had written tests in Danish, mathematics, and German. We could also sign up for other interesting subjects: photography, classical music, and needlework. In Danish, you were allocated to either group 1, 2, or 3, depending on how good you were. I wondered if I would be in group 1 – in Greenland, it was almost always the case that the broadcasters' children were in the top groups for Danish – and I worried about the possible shame of being placed in the bottom group, on the grounds of my being a Greenlander. But mine turned out to be amongst the best of the results. The tests had been just like the ones at school, where you had to answer oral questions about a piece of writing that you had done. I was also in the advanced group for German, and Ove was in one of the other groups. He wasn't really interested in German, and he'd signed up for it, in the hope of spending more time with me. The good thing about signing up for German classes was that you got to go on a trip to southern Germany when you finished the college programme. Ove also signed up for classical music, when he saw that I had done so.

Three weeks into our time there, we had our first big party in the gym. Ove and some of the others were on dishwashing duties, and they worked really quickly, because Ove was very keen to come and sit next to me, which he did. We had a really nice evening, which ended with us kissing on the stairs. There was a group of young men standing at the entrance to the gym, and we heard one of them ask, 'Who the hell is that, kissing Helene?' We heard another of them reply, 'It's him, Thiesen'.

After that party, Ove and I became inseparable. In the dining room, and in lectures, I never had to worry about being able to find a space to sit. Everyone knew not to sit down next to Ove, because that was Helene's chair. Our love blossomed; we became each other's confidantes, and we couldn't bear to be apart. We told each other things about our life histories – we had both lost our fathers. I was often upset, feeling incredibly vulnerable, and suffering from homesickness, and Ove was just about to give up on me. On one occasion, I called him 'young man', and he promptly replied, 'I'm not a young man at all, I'm just a kid'. He was just twenty years old. 'How old are you?' he asked. I wouldn't tell him, because if he described himself as a kid, then what did that make me? He kept asking me about my age, though, and one time he noticed my passport in my room and grabbed it, looked at my date of birth, and quickly figured out how old I was. Then, he threw the passport on the table and left. 'Well, that's it for us, then', I thought. Ove had gone back to his room and had laid down on his bed, looking straight up at the ceiling, thinking. He had thought that I was eighteen, but I was twenty-three. He couldn't have known that I had been in the care system, and had been an institutional child until I was twenty-two, so that my educational stage didn't correspond with my actual age. Ove came back after twenty minutes. He knocked on the door, gently, and as soon as I opened it, he declared his great love for me. Age didn't matter. And from that point onwards, Ove has always built me up and given me the utmost love and faith in myself. After all, I didn't even think I was anything. After all, I was only a Greenlander, and an orphanage child, and I heard the old refrain at the orphanage many times: 'Don't ever think that you're anything'. Ove has always argued the opposite.

Many of us at the college had applied to train as educators. On the day that we were being informed as to whether or not we would be able to commence training as kindergarten teachers, I received the joyous news from Aalborg Kindergarten Teaching College that I could start there on September 1, 1968. I was really happy, and I was just about to start celebrating when I saw many of the others sitting there, crying. Out of seventy-eight applicants from Ry, only three of us had been admitted.

Until I formally graduated, the Ministry of Greenland took us in hand. They had found a place in a house for me, and I would be spending four months in Aarhus. Ove and I arranged, through the college, to get a room each with a lady in Egå.[2] In total, she rented rooms out to five people. We each had our own small kitchen cupboard, and we had a shared bathroom. Ove got a room on the second floor, and mine was on the ground floor. Whenever we went up or down the stairs, people would keep an eye on whether or not we returned to our own rooms. My first job there was with a company director in Risskov, whose wife was fanatical about cleaning. She would follow me around the place, wetting her index finger, and running it along the window sill, saying, 'There's still dust here!' I had to eat alone in the kitchen, and I got margarine on my bread, whilst the family got butter. I even had to iron the diapers. Ove sent me letters when he was in Nakskov, with his mother, and the director's wife hid these from me until it was my day off. Fortunately, I found myself a new job in an Aarhus home, where they had two lovely kids. The lady there didn't think that I should be making two sets of dinners, so Ove was invited to eat with us.

Our great love bore fruit – Laila was born on April 18, 1969, three days before my twenty-fifth birthday. In June 1970, I qualified as an educator from the Aalborg Kindergarten Training College. Ove had trained as a bookseller, but during his time as a guard at Amalienborg, he had decided that after his military service,[3] he too would train as an educator. So as I was finishing my training, he was starting his, at the Nykøbing Falster Kindergarten Training College. We agreed that we would work in Greenland when Ove finished training. We both got jobs as kindergarten teachers in Maniitsoq, where our daughter Anja was born, on the most beautiful snowy day, on November 28, 1972.

When I was at school in Nuuk, I bought a vinyl record called 'Someone, someone, someone who really loves you'.[4] Back then, I thought deeply about whether I would ever find someone who truly loved me. I also thought about if I did have my very own family, then my husband and my children would have all the love and kisses that I had longed for in my own childhood. My dream was to have a happy family.

When Laila was in the second grade, we moved to Denmark, because conditions were not optimal in the Greenlandic school – for example, there were new teachers there every year, and there were so few children in the Danish classes that they were often merged with the grades above or below. We had lived happily in Maniitsoq for five years – we had some good colleagues there, and some lovely kids in our classes, and we saw my Aunt Karoline, Uncle Samson, my lovely cousins and many other members of my mother's family regularly. But we wanted our children to get a good education, and they both got good graduation diplomas from Vordingborg Gymnasium. They have both been influenced by the environment in which they grew up – one is a kindergarten teacher, and the other is a school teacher.

My dream came true. We have always kissed and hugged a lot in my family. One of Laila's first boyfriends said, 'When you get to the Thiesens' house, there's a sign up saying, "Kiss, kiss, kiss"!'. Ove and I both became recreation leaders in our district. I've also held positions as an assistant guardian, a guardian and patient advisor at psychiatric hospital in Vordingborg. My last job was as a head of education at Brøderup Children's Centre; I taught for one half of the working day and undertook my I managerial duties during the other half. It felt lucky that the last place I worked at before I retired was the best workplace of my life. There was also the fact that I had come home again – my workplace was nextdoor to Brøderup primary school, which I had attended for seven years as a foster child.

We have been fortunate in gaining two good and kind sons-in-law. And the happiness of our lives has been topped off with our wonderful grandchildren: Marius, Amalie, Nele, and Ina.

Notes

1 Skovlunde is a town about eight miles west of central Copenhagen, and Herlev is a suburb of Copenhagen about six miles northwest of the city centre.

2 This is a suburb of the city of Aarhus.
3 Since 1849, under the Danish Constitution, a period of military service (which lasts between four and twelve months) has been mandatory for every physically fit male over the age of eighteen. (Women can do this voluntarily, but are not obliged to do so.) Amalienborg Palace has been the home of the Danish royal family since 1794.
4 This was recorded by the British band Brian Poole and the Tremeloes, and was first released in 1964.

THANKS

I would like to thank Lene Therkildsen, Jørgen Chemnitz, and Louise Friedberg, who believed in my book; Fatuma Ali who gave me refuge in Italy; my aunts, Karoline Heilmann and Sofiaraaq Holm; my Danish foster-sisters, Tove Bard-thrum and Vagn Greve; and my beloved husband, Ove Thiesen, who has been a fantastic support throughout the writing of this manuscript – we have both cried and laughed together.

ABOUT HELENE THIESEN

Helene was born in Nuuk in 1944. She received her high school diploma from Ilinniarfissuaq College in Nuuk in 1965. In the winter of 1967–1968, she was at Ry University College, where she met her husband, Ove. She qualified as a kindergarten teacher in 1970. From 1972 to 1977, she worked as a kindergarten class leader in Maniitsoq, and in 1977 she was appointed as the leader of the Mern Recreational Centre in Langebæk district, where she worked for twenty years. From 1997 to 2006, she worked as a pedagogue in the Brøderup Children's Centre at Tappernøje. Soon afterwards, she and her family moved to Stensved in Sydsjælland, where she was invited to teach Danish to the Greenlandic patients at the state hospital in Vordingborg. That quickly led to other positions, first as a visitor to Greenlandic psychiatric patients, and then later as an assistant and supervisory guardian for criminal and mentally ill Greenlanders.

Helene first came into contact with the author, Tine Bryld, in connection with Bryld's work on the book, '*De nederste i Herstedvester*' ['The Lowest in Herstedvester (Prison)'], which was about Greenlanders who were serving prison sentences in Denmark. Helene told her own story to Tine Bryld, and only then became aware that she and her 'fate siblings' had been the subjects in a social and human experiment. Helene wrote her own section in Tine Bryld's book, '*I den bedste mening*' ['With the Best of Intentions'] (Figures 37 and 38).

In 2006, Helene took early retirement and undertook an after-school course called, 'Write your life'. That's how she really got started on writing her own story. In 2007, Helene was hired as a consultant for the film project. '*Eksperimentet*' ['The Experiment'], which had its premiere in Nuuk in September, 2010. Helene's own book, '*For flid og god opførsel: vidnesbyrd fra et eksperiment*' ['For Diligence and Good Behaviour: Testimony from an Experiment'] was published in 2011. Helene has given many lectures since the release of '*I den bedste mening*' and '*For flid og god opførsel*'.

FIGURE 37   Researching for 'I den bedste mening' – Helene and an iceberg (1998)

FIGURE 38   Helene at the launch of 'I den bedste mening' (August 1998)

FIGURE 39 Helene Thiesen in August 2016

TRANSLATOR'S AFTERWORD

Stephen James Minton

At the time of this book going to press (March 2022), we are at a distance of only a few weeks concerning some significant developments (that is to say, an award of financial compensation, and face-to-face apologies on behalf of the State from a currently serving prime minister) regarding actions that occurred more than a Biblical human lifetime (three-score-and-ten years) ago. It is possible to say, as I did in my introductory chapter, that the face-to-face apologies appeared to have meant something significant to the survivors, at the time that they were made, and in the days immediately thereafter. Unfortunately, and due in no small part due to their recency, it is not possible to predict what the medium- to longer-term effects of those actions might be with any full measure of confidence.

It is, of course, important to consider what was not – and could not have been – restituted in the compensation awards and face-to-face apologies of February and March 2022. The most glaringly obvious feature is that of the twenty-two Inuit children who were taken away from their homes and families in 1951, only six remain alive today, and that these people are all in their late seventies. Hence, sixteen of them did not survive long enough to hear these apologies, and every indication is that their adult lives, as well as their childhoods, were exceptionally difficult. We also have to acknowledge that the face-to-face apologies in March 2022 did not occur in a spontaneous act of Danish governmental conscience, nor in the context of a greater Danish awareness of the colonial relationship between Denmark and Greenland, and the abuses committed by Denmark therein. As we have seen in my introductory chapter, Mette Frederiksen's recent actions as prime minister in this case have been exceptional *because* they have occurred after seven decades of denial and obfuscation by the Danish government and authorities. It is noteworthy, of course, that the disruption of this denial and obfuscation has been mounted chiefly by those who survived the 'experiment' themselves, their families and friends, and the journalists, authors, and handful of politicians who have allied themselves to this cause.

The award of compensation – again, made to the six survivors of the 'experiment' who are still living – was hard won. The very human response to the amounts awarded (which was DKK 250,000; ca. £28,000/€33,500/CA$47,000/US$37,000 to each) is to consider them in the contexts of what is now known about the experiences endured by those who went through the 'experiment' and its seventy-year aftermath. What price should be placed on things like a stolen childhood, the deliberate stripping of individual and cultural identity and their markers, the devastating consequences to family life, the compromised adult opportunities, and the experiences of obfuscation and cover-up? (What price would you, dear reader, place on your childhood, your family, and your identity?) Whilst the law and the calculation of compensatory awards often seem, to the layman, to follow their own peculiar course, I do not think that we should fully dismiss these layman's/human responses. What if the compensation awards and face-to-face apologies of February and March 2022 were to mark an end point in the quest for restitution and justice for the survivors of the 'experiment', rather than a starting point? Could we then expect future historical commentators to be satisfied that the state of Denmark had discharged its responsibilities for violating the human rights of twenty-two children for a financial outlay of a mere DKK 1.5 million (ca. £168,000/€201,000/CA$282,000/US$222,000)?

Helene has commented on never having been able to fully rebuild her relationship with her mother, and we have seen in her text that the Danification measures implemented in the orphanage were, at the very least in terms of her ability to speak her native tongue, highly effective. Helene records that she had to work hard to learn Greenlandic as a young adult, which was also when she had her first opportunities to spend any real time with her immediate family and her extended family networks. It is always worth recalling that Helene first found out that she had been involved in an 'experiment' when she was in her forties. Elsewhere, we have considered this fact in evidencing the extent and longevity of the secrecy that surrounded the 'experiment', but let's also think about it in the context of the family that Helene was able to found with her beloved husband, Ove. The revelation coincided with the time that her own daughters, Laila and Anja, were entering their own adulthood. Laila and Anja have raised children of their own, and those children have now passed the age at which their grandmother was, when she was taken away from Greenland. For all of their adult and child-rearing lives, Laila and Anja have seen their mother struggle for recognition of, and some sense of justice and peace with, what happened to her as a child. It is impossible to think that the 'experiment' did not have intergenerational effects. Whilst Indigenous authors and peoples in North America and Australasia have had a good deal to say about the transmission of intergenerational trauma (concerning many aspects of colonialism and assimilation efforts, including the experience of residential schooling – see, for example, Bougie & Senecal, 2010; Brave Heart & DeBruyn, 1998; Cooper & Driedger 2019; Feir,

2016; Marsh, Marsh & Najavits, 2020; Wesley-Esquimaux, 2007; Winder, 2020), we in Europe have not previously addressed similar realities to nearly the same extent (Juutilainen, Miller, Heikkilä & Rautio, 2014). Suffice it to say, that whilst the number of survivors of the 'experiment' is inevitably reducing year-on-year, it is highly probable that the intergenerational effects of it will continue to be lived out. A recent review of the research literature indicates that this has been a serious problem in other country contexts (see Panofsky, Buchanan, John & Goodwill, 2021). Likewise, as the truth commissions elsewhere in the Nordic countries engage with the lived realities of the assimilation efforts made against the Sámi and Kven peoples in Norway, Sweden, and Finland, the intergenerational effects of measures such as forcible residential schooling will surely become more apparent. It is clear that we will need to find ways to conceptualise, talk about, and address intergenerational trauma amongst Indigenous peoples of the Nordic countries, and the lands to which those modern nation states have laid claim.

Over recent decades, various types of expressions and statements of apology to Indigenous peoples have been made by various church and state representatives around the world. The political scientist Matt James (2008) has offered a scheme by which the degree of authenticity (or non-authenticity) of such apologies can be ascertained.[1] In my view, Mette Frederiksen's face-to-face apologies to the survivors of the 'experiment' satisfy the James criteria, and to give Frederiksen the full credit that she is due, her spending days in both Denmark and Greenland with the remaining survivors and their families would seem to considerably exceed what has been done by prime ministers in similar situations elsewhere. I would be happy to go as far as to say that Frederiksen's actions offer a model for how her non-Danish counterparts might proceed in future. That being said, in her review of settler-state apologies to Indigenous peoples, Sheryl Lightfoot (2015, p. 17) made the following point:

> A state apology to Indigenous peoples must meet two criteria in order to be meaningful…It must, first, fully and comprehensively acknowledge the wrongs of the past and/or the present. Second, the state must make a credible commitment to do things differently, to make substantial changes in its policy behaviour, in the future. Any state apology that fails to deliver both of these two elements will not be meaningful in the eyes of Indigenous recipients, regardless of how 'authentically' it is delivered by the state, as judged by the James criteria.
>
> *(p. 17)*

Hence, to return to the metaphor from Malik el-Shabazz that I utilised in my introductory chapter in considering the points that I have raised above, more needs to be done in order to heal the wound that the knife in the back has made. I would like to reiterate my sense that the compensation awards and face-to-face apologies of February and March 2022 should mark a starting point, rather than an end point, in addressing the case for restitution and justice for the survivors of the 'experiment'. Importantly, what is distinctly less clear at present is what has been, or might be, offered to the families of the sixteen survivors of the 'experiment' who are no longer still living, or to the thousands of Inuit people affected by Danish colonisation measures other than the 'experiment'. Additionally, the question of what changes might occur in the relationship between Greenlanders and Denmark at broader levels, as a result of these recent developments, remains outstanding.

At the end of this book, I would like to express how truly grateful and blessed I feel that, through our working together, I have come to know Helene Thiesen as a friend. Putting aside the observations, speculations, and tentative and direct recommendations that I have provided in this afterword (expected as they probably are, from an academic like me), what I most hope for now is that my friend, and her family, will finally be able to find some peace. And if, by playing my part in bringing Helene's story to English-speaking audiences, some small but meaningful contribution can somehow be made to enhancing the possibilities for justice and accountability in other situations of the genocides and abuses committed against Indigenous peoples, then that would be more than I could ever have hoped for.

Helene and I wish to thank you for reading this book.

References

Bougie, E. & Senecal, S. (2010). Registered Indian children's school success and inter-generational effects of residential schooling in Canada. *International Indigenous Policy Journal*, 1(1): 33–60.

Brave Heart, M.Y.H. & DeBruyn, L.M. (1998). The American Indian holocaust: Healing historical unresolved grief. *American Indian and Alaska Native Mental Health Research: Journal of the National Center*, 8(2): 56–78.

Cooper, E. & Driedger, M. (2019). 'If you fall down, you get back up': Creating a space for testimony and witnessing by urban Indigenous women and girls. *International Indigenous Policy Journal*, 10(1): 1–22.

Feir, D. (2016). The intergenerational effects of residential schools on children's educational experiences in Ontario and Canada's western provinces. *International Indigenous Policy Journal*, 7(3): 1–44.

James, M. (2008). Wrestling with the past: Apologies, quasi-apologies, and non-apologies in Canada. In M. Gibney, R. Howard-Hassman, J.-M. Coicaud & N. Steiner (eds), *The Age of Apology: The West Faces Its Own Past*. Philadelphia: University of Pennsylvania Press, pp. 137–153.

Juutilainen, S.A.; Miller, R.; Heikkilä, L. & Rautio, A. (2014). Structural racism and Indigenous health – What do Indigenous perspectives of residential school and boarding school tell us? A case study of Canada and Finland. *International Indigenous Policy Journal*, 5(3): 1–18.

Lightfoot, S. (2015). Settler-state apologies to Indigenous peoples: A normative framework and comparative assessment. *Native American and Indigenous Studies*, 2(1): 15–39.

Marsh, T.N.; Marsh, D.C. & Najavits, L.M. (2020). The impact of training Indigenous facilitators for a Two-Eyed Seeing research treatment intervention for intergenerational trauma and addiction. *International Indigenous Policy Journal*, 11(4): 1–20.

Panofsky, S.; Buchanan, M.J.; John, R. & Goodwill, A. (2021). Indigenous trauma intervention research in Canada: A narrative literature review. *International Indigenous Policy Journal*, 12(2): 1–24.

Wesley-Esquimaux, C.C. (2007). The intergenerational transmission of historic trauma and grief. *Indigenous Affairs*, 4/07: 6–11.

Winder, N.N. (2020). Colliding heartwork: The space where our hearts meet and collide to process the boarding school experience. In S.J. Minton (ed), *Residential School Systems and Indigenous Peoples: From Genocide via Education to the Possibilities for Truth, Restitution, Reconciliation and Reclamation*. London: Routledge, pp. 141–162.

Note

1 In brief, these criteria are (i) recording the apology officially in writing; (ii) naming the wrongs; (iii) accepting responsibility; (iv) stating regret; (v) promising non-repetition; (vi) not demanding forgiveness; (vii) non-arbitrariness; and (viii) undertaking, through ceremony or reparation, efforts to indicate sincerity. In this model, in order to be ascertained 'authentic', an apology must satisfy all of these criteria; those which are not 'quasi-apologies' or 'non-apologies'.

INDEX

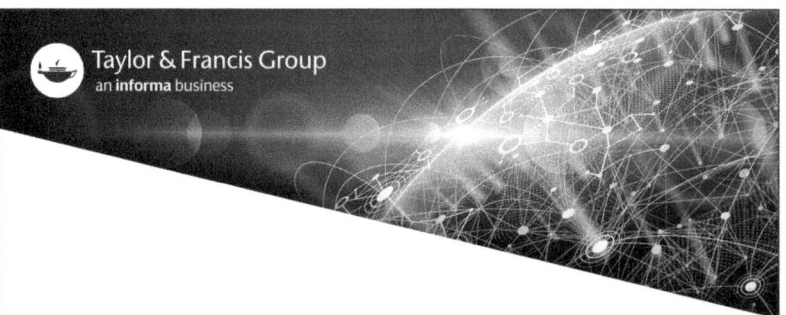